ANCIENT ASTRONOMICAL OBSERVATIONS
AND THE ACCELERATIONS
OF THE EARTH AND MOON

by Robert R. Newton

THE JOHNS HOPKINS PRESS
BALTIMORE AND LONDON

CONTENTS

FIGURES   vii
TABLES   ix
PREFACE   xiii

PART ONE: THE ANCIENT OBSERVATIONS

I — INTRODUCTION   1

II — OBSERVATIONS OF SOLAR AND LUNAR
     POSITION   9

    1. Hipparchus' Equinox Observations   9
    2. Ptolemy's Equinox and Solstice
        Observations   17
    3. Islamic Equinox and Solstice
        Observations   25
    4. Miscellaneous Observations of Solar
        and Lunar Position   29
    5. Summary of the Results of Chapter II   34

III — REMARKS ON SOLAR ECLIPSES AND
      ANCIENT ECLIPSE RECORDS   35

    1. General Comments about Total Solar
        Eclipses and Records   35
    2. Classes of Eclipse Reports and Their
        Reliability; or, How to Believe Six
        Impossible Things before Breakfast   43

IV — ANCIENT RECORDS OF LARGE SOLAR
     ECLIPSES   49

    1. Eclipse Reports from the British Isles   49
    2. Eclipse Reports from Babylonia and
        Assyria   57
    3. Eclipse Reports from China   61
    4. Eclipse Reports from Europe   70
    5. Eclipse Reports from Mediterranean
        Countries   91
    6. Eclipses That Will Not Be Calculated   121

iii

CONTENTS (continued)

V — MEASUREMENTS OF ECLIPSE MAGNITUDES
AND TIMES   123

   1.   Comments on Measurements of Time
       and Magnitude   123
   2.   Babylonian Measurements of Eclipse
       Magnitudes and Times   135
   3.   Chinese Measurements of Eclipse
       Magnitudes and Times   143
   4.   Islamic Measurements of Eclipse
       Magnitudes and Times   145
   5.   Mediterranean (Greek) Measurements
       of Eclipse Magnitudes and Times   152

VI — CONJUNCTIONS AND OCCULTATIONS   155

   1.   Stars Appearing in the Records of
       Conjunctions and Occultations   155
   2.   Conjunctions and Occultations
       Preserved by Ptolemy   156
   3.   Conjunctions and Occultations
       Preserved by Ebn Iounis   164

REFERENCES FOR PART ONE   175

PART TWO: THE ACCELERATIONS

VII — INTRODUCTION TO PART TWO   185

VIII — ANALYSIS OF THE CONJUNCTIONS AND
OCCULTATIONS   189

   1.   Computation of Stellar Coordinates   189
   2.   Lunar Conjunctions and Occultations   191
   3.   Planetary Conjunctions and Occulta-
       tions   200

IX — ANALYSIS OF THE MAGNITUDES OF
PARTIAL LUNAR ECLIPSES   211

   1.   General Comments about Magnitudes
       of Lunar Eclipses   211

iv

CONTENTS (continued)

2. Deductions from the Observed
   Magnitudes   214
3. Summary of Results from Lunar
   Eclipse Magnitudes   221

X — ANALYSIS OF LUNAR ECLIPSE TIMES   223

1. Preliminary Considerations   223
2. The General Procedure Used in the
   Analysis   225
3. Summary of the Results   228
4. Discussion   235
5. A Digression on Keeping Unequal
   Hours by the Stars   236

XI — ANALYSIS OF THE MAGNITUDES OF
     PARTIAL SOLAR ECLIPSES   239

1. General Comments about the Analysis   239
2. Summary of the Calculated Results   241

XII — ANALYSIS OF THE TIMES OF SOLAR
      ECLIPSES   245

1. General Comments on the Analysis   245
2. Summary of the Calculated Results   247

XIII — ANALYSIS OF THE LARGE SOLAR
       ECLIPSES   251

1. The Methods of Analysis   251
2. Records with Ambiguous Values of
   $\dot{\omega}_e$: Reassessment of Multiple
   Identifications   258
3. Summary of the Results   264

XIV — THE ACCELERATIONS AND THEIR
      VARIATIONS IN TIME   267

1. Recapitulation of the Material Available
   for Estimating the Accelerations   267
2. The Method of Inference   269
3. The Inferred Accelerations and Their
   Stability   272

CONTENTS (continued)

4. Comparison with Fotheringham's Results   275

5. A Probable Time History of the Accelerations   278

6. Discussion; Some Geophysical Speculations   286

REFERENCES FOR PART TWO   295

INDEX   301

FIGURES

V. 1 Schematic Representation of the Assumed
      Meaning of the Beginning and Ending
      Phases of Eclipses   124

V. 2 Ptolemy's Relation between the Magnitude
      of the Diameter and the Magnitude of the
      Area for Solar and Lunar Eclipses   130

VIII. 1 Calculated Positions of the Moon and of
         $8\beta^1$ Sco on -294 Dec 21   192

VIII. 2 Calculated Positions of the Moon and of
         Spica on -293 Mar 9   193

VIII. 3 Calculated Positions of the Moon and of
         the Pleiades on -282 Jan 29   193

VIII. 4 Calculated Positions of the Moon and of
         Spica on -282 Nov 9   194

VIII. 5 Calculated Positions of the Moon and of
         the Pleiades on 92 Nov 29   194

VIII. 6 Calculated Positions of the Moon and of
         Spica on 98 Jan 11   195

VIII. 7 Calculated Configuration of the Moon and
         of Part of the Scorpion on 98 Jan 14   196

VIII. 8 Longitude Error in Reported Conjunctions
         of Venus as a Function of Separation in
         Latitude   210

IX. 1 Differences between Observed and Calcu-
       lated Lunar Eclipse Magnitudes   217

IX. 2 Displacement of the Node Needed to Make
       Observed and Calculated Lunar Eclipse
       Magnitudes Agree   218

X. 1 Δt Derived from Observed Times of
      Lunar Eclipses   230

X. 2   Values of the Acceleration Parameter
       Derived from Lunar Eclipse Times, as a
       Function of Time   234

XI. 1 The Central Line of the Total Eclipse of
       866 Jun 16   241

XIII. 1 Geometry of the Observations of the
        Solar Eclipse of -309 Aug 15   251

XIII. 2 Contribution of Uncertainty in Observer's
        Position to Uncertainty in Ephemeris
        Longitude   253

XIII. 3 Contribution of Uncertainty in Observed
        Magnitude to Uncertainty in Ephemeris
        Longitude   255

XIII. 4 Contribution of Path Width to Uncertainty
        in Ephemeris Longitude   256

XIII. 5 An Alternate Method of Inferring Parame-
        ters from a Solar Eclipse   257

XIII. 6 Geometry of Two Discarded Possibilities
        for the Eclipse of Hipparchus   261

XIV. 1 Estimated Values of the Accelerations
       Plotted as Functions of Time   280

XIV. 2 The Parameter $10^9(\dot{\omega}_e/\omega_e) - 0.622\dot{n}_M$
       Plotted as a Function of Time   283

TABLES

II. 1	The Equinox Observations of Hipparchus	9
II. 2	The Equinox and Solstice Observations of Ptolemy	17
II. 3	Islamic Equinox and Solstice Observations	24
II. 4	Miscellaneous Solstice Observations	30
IV. 1	Large Solar Eclipses Reported from the British Isles	49
IV. 2	Large Solar Eclipses Reported from Babylonia and Assyria	57
IV. 3	Large Solar Eclipses Reported from China	62
IV. 4	Large Solar Eclipses Reported from Europe	71
IV. 5	Large Solar Eclipses Reported from Mediterranean Countries	91
IV. 6	Eclipse Reports That Will Not Be Used Further	121
V. 1	Ptolemy's Table for Converting Eclipse Magnitudes from Digits of Diameter to Digits of Area	129
V. 2	Babylonian Measurements of Times and Magnitudes of Lunar Eclipses	136
V. 3	Chinese Measurements of Eclipse Magnitudes	144
V. 4	Islamic Measurements of Eclipse Times and Magnitudes	146-147
V. 5	Mediterranean (Greek) Measurements of Eclipse Times and Magnitudes	152
VI. 1	Stars Used in Records of Conjunctions and Occultations	155

VI. 2   Conjunctions and Occultations Preserved by Ptolemy   158

VI. 3   Islamic Records of Conjunctions or Occultations Involving Venus   172

VIII. 1   Analysis of Lunar Conjunctions and Occultations Preserved by Ptolemy   197

VIII. 2   Calculated Latitude Differences When the Observed Latitude Difference Was Zero   204

VIII. 3   Acceleration of the Earth's Spin as Inferred from the Islamic Records of Conjunctions and Occultations   205

IX. 1   Magnitudes of Lunar Eclipses Observed before 500   215

IX. 2   Magnitudes of Lunar Eclipses Observed after 500   215

IX. 3   Summary of Results from Lunar Eclipse Magnitudes   221

X. 1   Times of Lunar Eclipses Observed before 500   228

X. 2   Times of Lunar Eclipses Observed after 500   229

X. 3   Summary of Results from Lunar Eclipse Times   233

XI. 1   Magnitudes of Solar Eclipses Measured after 500   242

XII. 1   Analysis of Solar Eclipse Times   247

XII. 2   Average Parameters for Solar Eclipse Times   248

XIII. 1 A List of Eclipse Records that Could Not Be Used because There Was No Way to Choose between the Branches of a Double-valued Function 259

XIII. 2 Analysis of Reports of Large Solar Eclipses before -400 262

XIII. 3 Analysis of Reports of Large Solar Eclipses between -400 and +60 263

XIII. 4 Analysis of Reports of Large Solar Eclipses between +60 and +500 263

XIII. 5 Analysis of Reports of Large Solar Eclipses after 500 264

XIV. 1 Material Available for Estimating the Accelerations Other than Magnitudes of Partial Solar Eclipses and Observations of Large Solar Eclipses 268

XIV. 2 Some Inferences of the Accelerations 271

XIV. 3 Fotheringham's Results and a Comparison with Them 275

XIV. 4 Values of the Variable $10^9(\dot{\omega}_e/\omega_e) - 0.622\dot{n}_M$ 282

PREFACE

The stately march of the stars across the night sky, and the motions of the seven classical planets among the stars, have inspired man's admiration since before the beginning of written history. His admiration has caused man to include astronomical observations in permanent records of all ages. The regularity of celestial motion has made these records valuable in at least two fields of research.

Celestial bodies move in accordance with the laws of motion and of gravitation, as modified by general relativity theory and perhaps by other refinements yet to be discovered. By combining physical theory with careful observations made during the past two or three centuries, astronomers can calculate the positions of celestial bodies at times in the past with remarkable confidence. Although the moon moves through about 5 million degrees of angle in a millenium, they can calculate the position of the moon two millenia ago with an error that is probably less than 2°.

Suppose that a recorded astronomical event is a fairly rare one, such as a large eclipse of the sun on a specified day of the year at a particular place, but suppose we do not know in advance how to relate the chronology of the record to our system of chronology. We first calculate the possible eclipses, and if they are far enough apart in time, we can say with assurance which eclipse was the one recorded. In this way we can then relate the ancient chronological system to our own. Thus ancient eclipses have aided our understanding of ancient history.

Celestial motions, however, are not exactly uniform. The slight variations in the rates of motion — the accelerations — are interesting to both the astronomer and the geophysicist. A comparison of ancient observations with what would be expected on the basis of modern astronomy alone gives us a powerful means of studying the accelerations; that is the main purpose of the present study.

The accelerations of the moon in its orbit and of the earth in its spin about its axis are the only accelerations large enough to study on the basis of current knowledge.

Of the many kinds of astronomical observations that will be used in this work, the one that occurs most often is that of a large eclipse of the sun. Study of ancient solar eclipses thus makes up a large part of this study.

Use of ancient observations to estimate the accelerations goes back about two centuries. It is disconcerting to find, however, that much of such work, including most of the work with solar eclipses, is wrong. The reasons for this remarkable situation are discussed at length in Chapter III, but the most important one is that many ancient references to eclipses are ambiguous, and many were almost surely not based on valid observations. Some references may have been magical; others may have been valid literary devices but not valid astronomical observations, to give two examples. It is necessary to decide which references to accept as observations, and it is sometimes necessary to decide between two possible interpretations of an ancient statement.

Virtually all studies of ancient eclipses that I know of have used the following procedure in handling doubtful or ambiguous cases: The author has assumed values of the accelerations in advance and has calculated the circumstances of the possible observations using them. He has then rejected as invalid all observations, or interpretations thereof, that do not agree well with the assumed values. He has finally used the remaining set of observations to calculate the accelerations. He necessarily found good agreement with his initial assumptions.

This is, of course, reasoning in a circle. I have used the term "identification game" to denote this particular example of circular reasoning (Chapter III). In an article[†] that was written while the present work was in

[†]Science, 166, pp. 825-831, November 1969.

press, I have described the workings of the identification game in more detail, and I have also summarized the quantitative conclusions of the present work.

The best way to avoid this logical fallacy is to make its occurrence impossible. In order to do so, I first studied the ancient records from the standpoint of the texts themselves, their historical settings, and other relevant considerations, before I did any astronomical calculations involving the observations. On the basis of the textual study, I adopted the astronomical interpretation that I would give the record. In order further to be sure that my interpretations were not influenced by agreement (or lack of agreement, as the case may be) with preconceived hypotheses, I wrote up all the interpretations as Part I of the present work and distributed Part I to a number of colleagues before I began the purely astronomical part of the study, which is the subject matter of Part II. That is the reason why this volume contains two parts.

Since writing Part I in its initial form, I have corrected typographical errors, and I have made changes intended to improve comprehensibility, or occasionally to strengthen a conclusion already reached. I have made a few substantive changes that are pointed out by notes. I have made no change that would affect the values of the astronomical accelerations derived from the observations.

References are listed separately for each part and are presented at the end of each part. Citations to the references are given by using the last name of the author (or the first author when there is more than one) underlined, followed by the year of publication enclosed within square brackets. Two or more works by the same author in the same year are distinguished by small letters after the year. When it is necessary to identify a particular part of a work, such as a section or a page, the needed information is placed within the brackets and after the year. Part of the title is used in place of the author's name for anonymous works.

I thank Mr. Clyde Holliday of the Applied Physics
Laboratory of The Johns Hopkins University (APL) for useful discussions about various visual problems involved in
the ancient eclipses. I thank Dr. R. P. Rich, also of APL,
for translating a number of Latin passages for me; I subsequently revised his translations to suit my own purposes
and absolve him of all responsibility. Professor Emeritus
R. E. Parker of the University of Tennessee gave Part I
a critical reading, helped with several translations that are
not specifically acknowledged, and discovered the eclipse
record called 1191 Jun 23 B. Professors H. W. Fuller,
Henry Kratz, and Albert Rapp, all of the University of
Tennessee, helped me find documents relating to the
alleged eclipses of Caesar and of Stiklestad. Mr. Sverre
Kongelbeck of APL obtained some of the Stiklestad docu-
mentation, translated it for me, and helped me prepare
the discussion relating to Stiklestad; responsibility for the
conclusions remains mine. Mr. Kaye Weedon of Blommenholm, Norway, found several items of Norse and German
origin and advised me on questions of usage. Mr. R. E.
Jenkins and Mr. I. H. Schroader of APL made a critical
review of many important calculations, and Dr. W. H.
Guier provided valuable advice on many difficult points.

My secretary, Mrs. Mary J. O'Neill of APL, deserves
special mention for her capable and patient assistance with
the preliminary drafts of this study.

I thank the library of the Naval Observatory for providing me with a number of hard-to-get documents and in
particular for allowing me to use the rare book cited as
Souciet [1729]. I also thank the reference staff of the APL
library for their patience and skill in locating many works
that are not normally the concern of a science librarian.

Part of the title of Section III. 2 is borrowed from
Dodgson [1872].

The staff of the Naval Observatory calculated the
circumstances of the solar and lunar eclipses, using the
standard ephemeris programs for the earth and moon. I

deeply appreciate their assistance with these eclipse calculations, which allowed me to verify the accuracy of the eclipse programs that I have prepared.

This work was supported by the Department of the Navy under its contract with The Johns Hopkins University.

PART ONE: THE ANCIENT OBSERVATIONS

CHAPTER I

INTRODUCTION TO PART I

Writers and astronomers of ancient times made many observations of astronomical phenomena. Some of their observations have been preserved; undoubtedly most of them have been lost. Use of ancient observations in attempts to deduce the secular acceleration[†] $\dot{n}_M$ of the moon's mean motion n_M began about 1750. <u>Newcomb</u> [1875] gives a summary, with references, of earlier work. <u>Ginzel</u> [1899] is almost encyclopaedic in his discussion of solar and lunar eclipses from the year -762 to 600, particularly from the standpoint of textual criticism of the eclipse records that he used.

The reader who is interested mainly in ancient solar eclipses may go directly to Section III.2. He should read Section III.2 before the eclipse reports themselves, since it deals with family resemblances between eclipse reports and suggests some principles that are useful in interpreting them.

Around 1900, use of the ancient records was enlarged to include attempts to find an "acceleration of the sun" as well as the lunar acceleration $\dot{n}_M$. There is a small acceleration term in the sun's (that is, in the earth's orbital) mean motion that comes from a gravitational theory, but this term is not a subject of study here. From the standpoint of the present work, the "acceleration of the sun" is

---

[†] There is an acceleration of the moon that results from planetary perturbations and that can be calculated from a strictly gravitational theory in which mechanical energy is conserved. The term "secular acceleration" of the moon and the notation $\dot{n}_M$, as they are used in this work, will not include the planetary contribution. They will mean the excess of the actual acceleration over that calculated from a strictly gravitational theory.

equivalent to the secular part $\dot{\omega}_e$ of the earth's spin acceleration[†]. If the earth's rotation is used to furnish the standard of time, the standard is called solar time. If the earth's rotation rate ω_e varies with respect to the stars, but if the rotation furnishes the time standard in spite of its variation, the sun appears to accelerate in its motion. "Acceleration of the sun" is a relic of the use of solar time as the astronomical standard.

The current astronomical standard of time is ephemeris time, which is the kind of time kept by the orbital motion of the planets. $\dot{n}_M$ and $\dot{\omega}_e$ will be with respect to ephemeris time; this point will be discussed in more detail below.

<u>Fotheringham</u> (in the series of papers through 1920 cited in the References for Part I) has made perhaps the most extensive analysis of ancient observations for the purpose of finding both $\dot{n}_M$ and $\dot{\omega}_e$.

Also during the twentieth century, efforts have been made to find current values of $\dot{n}_M$ and the spin acceleration from recent observations. At first these efforts were frustrated by the sporadic component of the earth's spin acceleration, which outweighs the secular component $\dot{\omega}_e$ over a span of a few centuries. <u>de Sitter</u> [1927] was perhaps the first to separate the lunar and earth accelerations by the use of solar and planetary longitudes in combination with lunar longitudes. <u>Spencer Jones</u> [1939] used this method to obtain the value $\dot{n}_M = -22''.44$ per century per century that is currently used in many national ephemerides. The same method yields accurate values for the earth's spin acceler-

---

[†]There are large sporadic accelerations, both positive and negative, in the earth's spin rate ω_e. In this work, $\dot{\omega}_e$ will mean only the "secular part" of this acceleration. It is not possible at present to give a rigorous definition of the secular part; it can only be considered as a particular average, to be defined in Section II. 1, since the time of an observation. It will not often be necessary to refer to the value of the acceleration at a particular epoch; when it is necessary, an appropriate circumlocution will be used.

ation as a function of recent time; it cannot yield the secular acceleration because of the short time span of the data.

Neither ancient nor modern observations yield strongly determined values of both $\dot{n}_M$ and $\dot{\omega}_e$, although the detailed reasons for difficulty are different. The difficulty with the ancient observations is shown by the spread of values of $\dot{n}_M$ and $\dot{\omega}_e$ obtained from them. Values obtained from different sets of ancient observations, or even from different interpretations of the same sets of observations, ranged over a factor of two. Also, the values of $\dot{n}_M$ finally adopted by Fotheringham [1920] and de Sitter [1927] are about 50 percent greater than the value found by Spencer Jones. These circumstances have given rise to two widely held but incompatible views: (a) the results from ancient observations are unreliable, and (b) the results from ancient and modern observations differ significantly.

Since neither the ancient nor the modern observations alone give reliable values for both $\dot{n}_M$ and $\dot{\omega}_e$, van der Waerden [1961], Curott [1966], and others combine them (see particularly the discussion, with numerous references, by Munk and MacDonald [1960]. They adopt the well-determined value of $\dot{n}_M$ from the modern observations and use it in analyzing the ancient observations in order to find a well-determined value of $\dot{\omega}_e$. The resulting pair of values rests on the explicit assumption that $\dot{n}_M$ has been constant since ancient times, and this method cannot contribute to the important question of whether there has been an appreciable change in $\dot{n}_M$ since then.

In this work I shall study the ancient observations for the purpose of answering two questions: (a) What are the best estimates of $\dot{n}_M$ and $\dot{\omega}_e$ that can be formed from the ancient observations? (b) In view of the uncertainties of observation, do the values of $\dot{n}_M$ from the ancient and the modern observations differ significantly? In the study I shall analyze again observations that have been used in earlier studies, and I shall use ancient observations that have not been analyzed previously.

I have arbitrarily chosen the solar eclipse of 1241 Oct 6[†] as the most recent observation that will be used. The word "ancient" in this context thus means "occurring before 1241 Oct 7".

The small amount of mathematical notation that is needed will be defined as it arises. A few preliminary remarks about the terms and notation used for time may lessen confusion in reading what follows.

Time appears both as an independent and as a dependent variable. Where it is the independent variable, it is always ephemeris time (ET). The word "century" means a Julian century of 36 525 ephemeris days, except where it is used loosely in phrases such as "the twentieth century". T denotes time measured in centuries from the fundamental epoch 1900 Jan 0.5 ET. JED means the Julian Ephemeris Date measured in ephemeris days; the fundamental epoch equals JED 2 415 020.0. "Century" will be abbreviated "cy".

Recorded times will be used for some ancient observations. It is assumed that the recorded time was some variety of solar time, and that this time can be related with sufficient accuracy to Greenwich mean solar time (GMST)[‡]. Calculations involving GMST will involve the Julian Date reckoned in mean solar days from 1900 Jan 0.5. The Julian

---

[†]Dates will be given in astronomical style, in which the year is followed by the month and then the day; 3-letter abbreviations without a period will be used for the months. Times shorter than a day may be given by hours, minutes, and seconds, or by means of decimal fractions of the day following the month.

[‡]GMST is often used in referring to sidereal time. Here "S" is inserted into GMT to indicate that solar time, not universal time, is needed. It is hoped that this will not cause too much confusion. The Explanatory Supplement [1961, Chap. 3] discusses the various kinds of time in detail.

Date reckoned in this way is denoted by JSD. Universal time (UT) is used only once in a trivial way in Section II. 1, where it is assumed that UT = GMST at the epoch 1900 Jan 0.5 GMST.

For identification purposes the calendar date of an ancient observation will be used. The year part of a date before the beginning of the common era is given in astronomical usage; in this usage, 0 = 1 B.C., -1 = 2 B.C., and so on. The dates of the ancient observations occur before the invention of the Gregorian calendar and are given in the reckoning according to the Julian calendar (not to be confused with the Julian Date). This is done even for dates before the introduction of the Julian calendar. In the dates the year is assumed to begin on Jan 1, regardless of any local custom that may have existed to the contrary at the time and place of an observation.

In a calendar date, the day is the day of ordinary usage that begins at midnight, and not the day of former astronomical usage that begins at noon. However, the day used in reckoning the JED or JSD does begin at noon.

In much earlier work involving the accelerations of the earth and moon, when GMST rather than ET was used as the time base, it was customary to derive values for the "accelerations of the sun and of the moon". Further, the word "acceleration" was often used loosely. It was often used to mean the coefficient of τ^2 in the longitude discrepancy of the sun or moon, where τ denotes GMST; thus there is an ambiguous factor of 2. For example, Fotheringham frequently used L for the longitude discrepancy of the moon and L' for that of the sun. He called L/τ^2 the acceleration of the moon and L'/τ^2 the acceleration of the sun, and measured both "accelerations" with respect to GMST. The relations that connect L/τ^2, L'/τ^2, $\dot{n}_M$ and $\dot{\omega}_e$ are

$$2L/\tau^2 = \dot{n}_M - 1.7373 \times 10^9 (\dot{\omega}_e/\omega_e),$$

$$2L'/\tau^2 = -0.12996 \times 10^9 (\dot{\omega}_e/\omega_e),$$

(I. 1)

if angles are measured in seconds and time in hundreds of years. Note that the acceleration of the moon with respect to GMST is not the same as $\dot{n}_M$.

One other difficulty may be encountered in reading some of the earlier work. Much of it was done before the size of the planetary contribution to the lunar acceleration was agreed upon, and varying amounts of it may be included in L/τ^2. In particular, in Fotheringham's papers, $6''.1/cy^2$ should be subtracted from his values of L/τ^2 before substitution into Eq. (I. 1).

Many national ephemerides currently use the values $n_M = -22''.44/cy^2$ and $\dot{\omega}_e = 0$; I shall call these the standard values. The first purpose of this work is to find the deviations from the standard values that best fit the ancient observations. If the ancient data have contributed significantly to the present determination of the ephemerides, use of a deviation from the standard accelerations would require corresponding adjustments in the appropriate terms that are linear in the time and that are independent of the time. It is quite difficult [Duncombe, 1968] to determine the quantitative contribution of ancient data to the standard values, and a separate investigation will be required on this point. In order to proceed with this study while waiting upon that investigation, it is necessary to make an arbitrary choice. I shall make the simplest choice, which is to take the changes in the longitude of the moon and in the right ascension of Greenwich to be simply $\frac{1}{2}\dot{n}_M T^2$ and $\frac{1}{2}\dot{\omega}_e T^2$, respectively. That is, the constant and linear terms will not be varied when the quadratic terms are varied.

I am responsible for translations in this work that are not specifically attributed; most translations from ancient languages are attributed. In dealing with ancient European languages, I have ventured to supply translations for some simple passages. I have also verified directly, and in some cases supplied, the translations from ancient European languages whenever it seemed likely that the choice of modern words would affect the astronomical implications significantly. There is risk in using translations by a layman in the languages concerned, but there is also risk in using a translation by an expert in the language who may not be primarily concerned with the astronomical implications of the modern words chosen. The latter risk is illustrated by the eclipse of Thales (see the record -584 May 28 M in Section IV. 5). The standard translation that will be cited uses the astronomical term "eclipse"; the Greek equivalent does not occur in the original.

CHAPTER II

OBSERVATIONS OF SOLAR AND LUNAR POSITION

1. HIPPARCHUS' EQUINOX OBSERVATIONS

Hipparchus' period of activity as a mathematician and astronomer extended from about -161 to -126; the dates of his birth and death are unknown. Most of his writing and most of his original astronomical observations have been lost. Ptolemy [ca 152; III. 2] has preserved 20 of his equinox observations. Fotheringham [1918] gives a translation of the relevant passages, and uses the observations to deduce a value of $\dot{\omega}_e$ (or of the acceleration of the sun).

Fotheringham reduced the times of the equinox passages of the sun observed by Hipparchus to GMST on the assumption that the recorded times were in apparent local solar time for the meridian of Rhodes. Hipparchus

TABLE II. 1

THE EQUINOX OBSERVATIONS OF HIPPARCHUS

Date	Greenwich Mean Time (hr)	JSD -1 600 000	JED -1 600 000	Δt^a (hr)
- 161 Sep 27	16	62 522.167	62 521.837	7.9
- 158 Sep 27	4	63 617.667	63 617.564	2.5
- 157 Sep 27	10	63 982.917	63 982.807	2.6
- 146 Sep 26	22	68 000.417	68 000.472	-1.3
- 145 Sep 27	4	68 365.667	68 365.714	-1.1
- 142 Sep 26	16	69 461.167	69 461.441	-6.6
- 145 Mar 24	4 1/4	68 178.677	68 178.990	-7.5
- 144 Mar 23	10 1/4	68 543.927	68 544.232	-7.3
- 143 Mar 23	16 1/4	68 909.177	68 909.474	-7.1
- 142 Mar 23	22 1/4	69 274.427	69 274.716	-6.9
- 141 Mar 24	4 1/4	69 639.677	69 639.959	-6.8
- 140 Mar 23	10 1/4	70 004.927	70 005.201	-6.6
- 134 Mar 23	22 1/4	72 196.427	72 196.655	-5.5
- 133 Mar 24	4 1/4	72 561.677	72 561.897	-5.3
- 132 Mar 23	10 1/4	72 926.927	72 927.140	-5.1
- 131 Mar 23	16 1/4	73 292.177	73 292.382	-4.9
- 130 Mar 23	22 1/4	73 657.427	73 657.624	-4.7
- 129 Mar 24	4 1/4	74 022.677	74 022.867	-4.6
- 128 Mar 23	10 1/4	74 387.927	74 388.109	-4.4
- 127 Mar 23	16 1/4	74 753.177	74 753.351	-4.2

[a] Δt = solar time - ephemeris time

was acquainted with the equation of time and could have used mean solar time; however, Fotheringham assumed that he did not, stating that the equation of time was rarely used by Ptolemy or earlier astronomers except in dealing with lunar observations. In any case it will appear that the difference between mean and apparent solar time is negligible in analyzing the observations. It is possible that some of the earlier observations were made at Alexandria rather than on Rhodes. The difference between solar time at the two places is about 8^m and is also negligible.

The dates and times of the observed equinox crossings, reduced by Fotheringham to GMST, are listed in the first two columns of Table II.1. Fotheringham used the day that begins at noon; I have changed his dates to the ordinary day. The JSD of each observation, in days and decimal fractions, is listed in the third column. 1 600 000 has been subtracted from the JSD for convenience in tabulating. The autumnal equinox observations are listed first, followed by the vernal ones.

There are two main sources of error in the observations. First, Hipparchus rounded the observed times to the nearest multiple of 6^h. The fact that 6^h is also nearly equal to the excess of the tropical year over 365 days produces curious effects in interpreting the observations, as Fotheringham has pointed out. Second, Hipparchus apparently determined the equinoxes by measuring the declination of the sun and interpolating to the times of zero declination. The possibility of an error in the setting of his equatorial plane must be considered.

Fotheringham used the observations listed in Table II.1 to estimate both $\dot{\omega}_e$ and the error δ_H in Hipparchus' equator. He then tested the value found for δ_H by comparing it with Hipparchus' values for the declinations of stars near the equator[†]. Since the equinox observations

---

[†] Because Hipparchus was ignorant of refraction, it is necessary to use only stars near the equator in estimating the

are rather slight to support the weight of estimating two parameters, it seems better to take δ_H from the stellar observations and to use this value in interpreting Table II.1.

From the stellar observations, Fotheringham finds

$$\delta_H = -0°.073 \pm 0°.018. \qquad (II.1)$$

The error quoted in Eq. (II.1) is the standard deviation, as all errors to be quoted in this work will be. The minus sign means that Hipparchus' declinations were too large and that the declination of the sun at the time of the observations was -0°.073 ± 0°.018 rather than zero. The table of Hipparchus' stellar positions [Fotheringham, 1918] also shows that the random error in Hipparchus' measurements of declination was 0°.045. This is enough to produce an error of about $2^h.7$ in the time of an equinox. However, Hipparchus must have used more than one measurement of the solar declination[†] in determining at least some of the equinox passages, since many of the quoted times occur at night. Hence his random error in declination in the equinox observations is probably less than 0°.045.

Table II.1 lists the JED for each equinox observation, with 1 600 000 again subtracted for convenience. The steps in the calculation of JED can be summarized briefly.

1. The declination bias in Eq. (II.1) means that the longitude of the true sun was -0°.182 at each vernal equinox and 180°.182 at each autumnal equinox. The nutation in longitude can be neglected, so the longitudes are taken as if they referred to the mean equinox of date.

---

error in his equatorial setting. This error is the appropriate one to apply to the equinox observations since the sun is on the equator at an equinox.

[†] If he worked as carefully as we hope he did. He could have found an equinox occurring at night by making a single daytime observation and calculating the time of zero declination.

11

2. The formulas on page 98 of the Explanatory Supplement [1961] (hereafter referred to as ES) lead to ω_S = 246°.242 and e_S = 0.0175 528 for the epoch T = -20.44; ω_S is the mean longitude of perigee and e_S is the eccentricity of the solar orbit. These values are used for all the observations. The error caused by using them instead of calculating ω_S and e_S separately for each observation is negligible.

3. The values found in steps 1 and 2 lead to the values L_S = -2°.035 at each vernal equinox observation and L_S = 182°.010 at each autumnal observation if we neglect perturbations whose periods are comparable with the year. L_S is the longitude of the mean sun.

4. In the formula for L_S as a function of T on page 98 of ES, the small quadratic term was evaluated for the epoch T = -20.44 and added to the constant term after the substitution T = T_1-20 was applied in double precision. This process yields

$$L_S = -95°.56078 - 360°(Y + 100) + 36000°.75682T_1 , \quad \text{(II. 2)}$$

where Y is the year part of the calendar date of an observation in Table II.1. Equation (II. 2) was then solved in single precision to give T_1 for each observation, and JED was found from

$$JED = 1\ 684\ 520 + 36525T_1 + 0.0058. \quad \text{(II. 3)}$$

The term $0^d.0058$ is the correction for light time.

The values of JED have been tested indirectly against the values in Nav. Obs. [1953]. The testing method will be described in the next section.

It is now necessary to form an estimate of $\dot{\omega}_e$ from each observation. By definition, GMST is related to T by

$$GMST = RA_G(T) - RA_{FMS}(T) + 12^h ,$$

$$RA_G(T) = RA_{Go} + \omega_e T + \tfrac{1}{2}\dot{\omega}_e T^2 , \quad \text{(II. 4)}$$

12

in appropriate units. $RA_G(T)$ is the right ascension of the Greenwich meridian, and $RA_{FMS}(T)$ is the right ascension of the fictitious mean sun, both taken with respect to the mean equinox of date. The constant RA_{Go} is implicitly defined by Eq. (II. 4). $RA_{FMS}(T)$ is given on page 74 of ES. Now consider the variety of time $(GMST)_o$ defined by

$$(GMST)_o = RA_{Go} + \omega_e T - RA_{FMS}(T) + 12^h.$$

It is clear that $GMST - (GMST)_o = \frac{1}{2}\dot{\omega}_e T^2$; hence this is the difference we want to form.

$(GMST)_o$ differs from ET for two reasons. First, GMST, UT, and $(GMST)_o$ all agree at the epoch 1900 Jan 0.5, but UT and ET disagree then by about 4^S (ES, Table 3.1); thus $(GMST)_o$ and ET disagree by about 4^S at $T = 0$. Second, RA_{FMS} contains the small term $0^S.0929T^2$, which amounts to about 40^S for the times of the observations in Table II.1. This dominates the difference at the fundamental epoch but is still negligible. Thus, to sufficient accuracy for this paper, $(GMST)_o$ can be equated to ET and hence

$$\dot{\omega}_e = (2/T^2)(JSD - JED) \qquad (II.5)$$

in appropriate units.

$RA_{Go} = 18^h.64589$ within an error that is much less than a second, and $\omega_e = 879\ 000^h.051\ 263\ cy^{-1}$.

The last column in Table II.1 lists, for each observation, the difference $\Delta t = JSD - JED$ converted to hours. Since the epochs of the observations are so close together, an average value of T^2 can be used in Eq. (II.5) for all observations. Thus it is sufficient to form a best estimate of Δt from the values listed in Table II.1, and then to apply Eq. (II.5) to this estimate.

The mean value of Δt from Table II.1 is $-3^h.8$. However, it is not legitimate to use the mean value in the peculiar circumstances that prevail here.

For one thing the first three observations do not belong to the same population as the others. They give the only positive values of Δt, and they are widely separated in time from the others. The most probable explanation is that Hipparchus improved the setting of his equatorial circle between -161 and -146, perhaps in two steps, one between -161 and -158 and the other after -157. It is possible that he moved from Alexandria to Rhodes between -157 and -146, so the later observations may have been made with a different instrument.

If this explanation is correct, the declination bias in Eq. (II.1) is the value to use with the observations after -146, since Hipparchus' stellar observations were made [Fotheringham, 1918] near the end of the series of equinox observations.

Even when the first three observations are deleted, a problem remains. It is connected with the fact that Hipparchus used 6^h as his minimum element of time in recording the equinox times.

In order to illustrate the problem, suppose for the moment that the tropical year equals 365^d 6^h exactly, and suppose further that the first vernal equinox observation in -145 occurred at 8^h local time on the date given. Then the equinox in successive years would have occurred at 14^h, 20^h, 2^h, and so on, and Hipparchus would have recorded all of them at the times that he did. In fact, his recorded times would have been unaltered for all the vernal equinoxes if the first one had occurred at any time between 3^h and 9^h. Under these circumstances a long series of observations, all compatible with each other, would have no more weight than a single one of them, and a statistical treatment would not be justified.

As it is, the tropical year is less than the Julian year by 11^m 14^s = $0^h.188$. Suppose that in some year the equinox fell just before 3^h so that it would have been assigned to 0^h by Hipparchus. The next year it would occur just before 9^h - $0^h.188$ and would have been assigned to 6^h,

and so on. The series of observations would go on for about
$6/0.188 \approx 32$ yr showing complete concordance with the
Julian year, and then would suddenly show a discordance of
6^h between two successive years. Thus, with the sparse set
of observations that we have in Table II. 1, the information
is to be found either from a discontinuity in the values of Δt,
or from the length of a series without a discontinuity.

The concordant vernal equinox series from -145 to
-127 furnishes two limits. If the discontinuity occurred
just before -145, the series would continue until -113, and
the value of Δt given by the entire series would be $-4^h.6$, the
value for the year -129. If the discontinuity occurred just
after -127, the value of Δt given by the series would be
$-7^h.1$, the value for -143. The best estimate for Δt that
can be formed from the vernal equinox observations alone
is the mean of these, namely $-5^h.8$. This is of course
close to the mean of the values from all the vernal equinox
observations. The point is that the mean is to be assigned
the weight of a single observation, not the weight of 14 observations.

There is a discontinuity in the values of Δt for the
autumnal equinoxes between -145 and -142. The vernal
equinoxes are given for each of these years, and for each
year in between, and show no discontinuity. The discontinuity thus does not seem due to any change in the observing procedure and is probably real. The best value to take
from the autumnal equinox records is therefore the mean
of the values of Δt for -145 and -142; this mean is $-3^h.8$.

It happens that the interval between the events that
Hipparchus called the vernal and autumnal equinoxes differs
from a multiple of 6^h by only about $0^h.5$, and a discontinuity in the vernal equinox observations would occur within
2 or 3 years from a discontinuity in the autumnal ones. A
discontinuity in the autumnal equinox between -145 and -142
thus implies that a discontinuity occurred in the vernal
equinox close to the beginning of the vernal equinox series.
Therefore, the value of Δt from the vernal series should be
$-4^h.6$, the first limit given above, and not $-5^h.8$, the mean
for the series.

15

Considering the autumnal and vernal equinoxes to-
gether, the best estimate of Δt that can be formed is $-4^h.2$.
I do not see any way to assign an error to this number on
the basis of the observations themselves. I shall arbitrarily assign a standard deviation of $1^h.0$, which seems
reasonable for the state of technology at the time.

The value of $10^9(\dot{\omega}_e/\omega_e)$ corresponding to $\Delta t =$
$-4^h.2 \pm 1^h.0$ is, by Eq. (II. 5),

$$10^9(\dot{\omega}_e/\omega_e) = -23 \pm 5 \ cy^{-1}. \tag{II. 6}$$

I propose this as the best estimate that can be formed from
the observations listed in Table II. 1. The "1σ values" from
Eq. (II. 6) are consistent with the values of Δt in Table II. 1
if consideration is given to the standard deviation of the
bias term δ_H.

The fact that the numerically greatest values of Δt
are about the same for both the vernal and autumnal equinoxes, after deleting the three early observations, confirms that Hipparchus found the equinoxes by observing
the declination of the sun, with a declination bias close to
that given in Eq. (II. 1).

Fotheringham [1918] found $10^9(\dot{\omega}_e/\omega_e) = -30 \pm 4 \ cy^{-1}$.
He recognized the problem posed by Hipparchus' method of
reporting the times and tried to meet it by using different
sets of observations. For most sets tried, however, he inferred a value for the declination bias and for $\dot{\omega}_e$ by a least
squares method; for one sample of the vernal observations,
he adopted the declination bias found from the stellar positions. He did not use the principal methods used here,
namely, the adoption throughout of the declination bias from
the independent stellar positions and the use of the discon-
tinuities as the main source of information.

Ptolemy [ca 152; III. 2] says that Hipparchus observed the summer solstice of -134 and that Aristarchus
observed the summer solstice of -279. Ptolemy gives the

interval between these two solstices but unfortunately does not give the observed time of either.

2. PTOLEMY'S EQUINOX AND SOLSTICE OBSERVATIONS

Ptolemy [ca 152; III. 2] gave the times of two autumnal equinoxes, one vernal equinox, and one summer solstice. Fotheringham [1918] studied the times of the equinoxes but not the time of the solstice.

TABLE II. 2

THE EQUINOX AND SOLSTICE OBSERVATIONS OF PTOLEMY

| Observed Time | | JSD | JED -1 700 000 | | Hour Calculated |
Date	Hour	-1 700 000	By Long.	By Decl.	From Older Tables
132 Sep 25	14	69 538.995	69 538.811	69 537.959	13.8
139 Sep 26	07	72 095.701	72 095.507	72 094.655	07.2
140 Mar 22	13	72 273.965	72 274.318	72 272.936	13.2
140 Jun 25	02	72 368.500	72 368.248	72 367.130	01.2$_a$ or 02.3

^{a}Depending upon the starting point used.

Table II. 2 lists the dates and the local (not GMST) times reported by Ptolemy. He reported the hour by giving the number of hours after one of the cardinal times of day. For example, he gave the hour of the 139 Sep 26 equinox as 1 hour after sunrise; this is listed as 07^h. Table II. 2 also gives the reported times converted to JSD. The JSD is, as usual, calculated on the assumption that the reported times are in apparent solar time for the meridian of Alexandria.

Many writers [Pannekoek, 1961, p. 149, for example] have commented that the reported times are wrong by a day and that the longitudes in Ptolemy's star catalog are wrong by about 1°. This suggests the hypothesis that Ptolemy, in his observations of the sun, was mainly interested in finding what we call the eccentricity and the position of perigee of the solar orbit. His raw material for finding these would be the lengths of the seasons, and

17

he would want the intervals between the epochs of the equi-noxes and solstices with as much precision as possible. Thus it would be reasonable to establish these epochs by finding when the longitude of the sun was a multiple of 90° rather than by using the more fundamental but less pre-cisely measured times when the declination of the sun was 0° or an extremum. On this basis he would have estab-lished an equinoctial position and referred all longitudes to it. A single error in the position of the equinox, which would not have affected his main purpose particularly, would then explain both the time errors in Table II. 2 and the longitude errors in the star catalog.

Peters and Knobel [1915] found that the mean error of the longitudes in Ptolemy's star catalog is -34'. 9 when referred to the equinox of 100. 0; the sign means that the tabulated longitudes are too small. Peters and Knobel do not state whether they used the mean or the true equinox. The mean epoch of the star catalog [Fotheringham, 1918] is 137. 55. The cumulated precession over 37. 55 yr is 37. 55 × 50". 3 ≈ 1889" ≈ 31'. 5, and the mean error in longi-tude with respect to the equinox of date is -66'. 4. I have joined Peters and Knobel in ignoring the nutation in this situation.

The column labelled "JED - 1 700 000, By Long. " gives the JED (less 1 700 000 for convenience) when the true longitude of the sun was 66'. 4 plus an appropriate multiple of 90°. The calculations were done by the method described in Section II. 1, except that the intermediary epoch $T = -18$ was used instead of $T = -20$. e_S and ω_S were calculated for the epoch $T = -17. 64$.

Another possibility is that Ptolemy found his equi-nox times by measuring the declination of the sun, just as Hipparchus almost surely did. This possibility requires the unlikely circumstance that Ptolemy's calendar was in error by one day. Accordingly, I also calculated JED for each equinox observation by the method of Section II. 1, us-ing -0°. 108 [Fotheringham, 1918] rather than -0°. 073 for the declination bias. For the solstice observation, I assumed that the longitude of the true sun was the mean of the true longitudes used for the equinoxes. These values of JED (diminished by 1 700 000) are listed in Table II. 2 in the column "JED - 1 700 000, By Decl. ".

The values of JED in Table II. 2 and those in Table II. 1 were tested against each other by calculating the values for Table II. 2 with the identical program used for Table II. 1 and vice versa. The negligible differences found in this way are attributable to the difference in the treatment of the nonlinear terms.

The values of JED in Table II. 2 were also tested against those found by inverse interpolation from Nav. Obs. [1953]. This document gives the longitude of the sun and the latitude and longitude of the moon, beginning from about -100; the solar table continues to about 1800 and the lunar table to about 2080. The angles are rounded to $0°. 01$. The largest discrepancy found was about $0^d. 01$. This is consistent with the rounding in the tables and with the errors introduced by neglecting the high-frequency terms in the solar ephemeris, and is negligible for present purposes. The values of JED in Table II. 1 cannot be tested against Nav. Obs. [1953] directly because they are too early; however, the combination of tests checks them indirectly.

Neither set of JED in Table II. 2 agrees well with JSD. The values of JSD - JED are positive rather than negative, except for the observation of 140 Mar 22 calculated by longitudes. The intervals between vernal and autumnal equinoxes calculated by longitudes are wrong by about 12^h, an impossibly large value. The values of JSD - JED calculated from the declinations remain positive even when the dates are adjusted by 1^d. The intervals between equinoxes calculated from the declinations, however, agree within better than $0^h. 5$.

The values of JSD - JED for the three equinoxes, calculated by the declination method, equal $24^h. 9$ with a maximum deviation from this value of $0^h. 25$. However, we can be sure that the correct value is about -3^h. Thus,

19

Ptolemy's times are about 28^h too large. It does not seem possible to account for a consistent error of this amount, even if we allow for a dating error of 1^d. For example, the equinox of 139 Sep 26 must have occurred at about 03^h on Sep 25, that is, at about 3^h before sunrise. Yet Ptolemy reported the event as happening "one hour approximately after sunrise".

I can think of only one plausible explanation of the times reported by Ptolemy. Hipparchus [Ptolemy, ca 152; III. 2] deduced that the length of the tropical year was 365$\frac{1}{4}$ days less 1/300 of a day, from a comparison of observations made by himself with still earlier observations. Hipparchus also assigned 178$\frac{1}{4}$ days to the interval from an autumnal equinox to the next vernal equinox, and 94$\frac{1}{2}$ days to the interval from then to the summer solstice. The local times of Ptolemy's equinoxes and solstices using these intervals and Hipparchus' length of the year, starting from Hipparchus' equinox measurement of -145 Sep 27, are listed in the last column of Table II. 2. In these calculations I have ignored the small difference between the meridians of Rhodes and Alexandria as Ptolemy would have done. I have also ignored the equation of time, which is equivalent to taking all times as local mean solar time.

In every case, the calculated date agrees with the date reported by Ptolemy. The hour of the day, when rounded to the integer hour, agrees exactly for each equinox. For the solstice, the calculated value is the first listed value, 01^h. 2, and the reported hour is 02^h.

The discrepancy in the solstice time may be considered as an argument against the proposed explanation. It may also be attributed to an error in calculation or copying. There is another and more probable explanation. Ptolemy [ca 152, Bk. 3] calculated the length of the year from the autumnal equinox observations in Table II. 2 and from the autumnal equinox observation of -146 from Table II. 1, and found that he agreed with Hipparchus' value of 365$\frac{1}{4}$ days less 1/300 of a day. Ptolemy also reported the summer solstice observation attributed to "pupils of Meton and Euctemon" in Section II. 4 below, and compared it with his own solstice observation in order to check the length found for the year. If 571 intervals of $[365 + (\frac{1}{4}) - (1/300)]^d$ are added to the time of the summer solstice of -431, the result is $02^h.3$ on the day stated in 140. When rounded to the even hour, this agrees exactly with the reported time.

It would take us too far from the purpose of this study to examine in detail all the suggestions that have been made in order to account for the errors in Ptolemy's times. They include an error in recording dates, an error in calculating the time interval between Hipparchus' and Ptolemy's observations, and a single bad observation (or recording) of the equinoctial position. None of these accounts for all the records.

Clerke [1910] attributes to Newcomb the finding of "palpable evidence that the discrepancies between the two series[†] were artificially reconciled on the basis of a year 6^m too long, adopted by Ptolemy on trust from his predecessor." Unfortunately, Clerke does not give the source of this finding. Since Clerke refers to Newcomb's use of Leverrier's solar tables, which are irrelevant to the question of how Ptolemy calculated, it is not clear how Newcomb reached the conclusion quoted.

If Ptolemy did calculate his equinox times, he would have started from the equinox of -146 or -145 and not from the later equinox of -142. He would have considered the last value to be erroneous since it seems to disagree with the self-consistent earlier equinoxes.

Ptolemy did not intend the times and other "data" to be used merely in exercises to illustrate his models of the solar system. For example, he said [Bk. III. 2] of his

---

[†]That is, the series of equinox observations of Hipparchus and Ptolemy.

observation of 139 that ". . we observed again with sureness the autumn equinox . ."[†], and he made similar remarks about other observations. Thus there is little question that Ptolemy intended the times to be accepted as genuine observations.

Ptolemy could have been the victim of a trusted colleague or assistant. Equally possible, he could have been responsible for using calculated rather than observed numbers. In either case we cannot judge the act in the same way that we would judge a similar act today; we do not know enough about the second century concepts of scientific "truth" and how it was sought.

Is there a connection between the solar errors and the errors in the star catalog? In 28^h the sun moves about 1°.15. The motion is in the right direction and has about the right size to account for the average longitude error for the stars. For this reason Fotheringham [1918] concluded that Ptolemy, having somehow made a large error in measuring the longitude of the sun at some epoch, later transferred the solar error to the stars by an understandable observational technique. This is possible whether the bad solar position came from a calculation or from bad observation. However, this conclusion does not account for the errors in the times of the equinoxes and the solstice, which should have been measured by following the declination, not the longitude, of the sun.

On the other hand, if we assume that Ptolemy or a colleague calculated the position of the sun from older tables, we can by this single assumption account for all the known errors in Ptolemy's work.

An alternate explanation of the errors in the star catalog is based upon Ptolemy's erroneous value of the precession of the equinox. Ptolemy [ca 152, VII. 2; also see Pannekoek, 1961, p. 126] took Hipparchus' original estimate of the precession, which was 2° in 169ʸ, first rounded

---

[†] "... ημεις ετηρησαμεν ασφαλεζατα παλιν την μετοπωρινην ισημεριαν ..." "Again" alludes to measurements of other equinoxes, not other measurements of this one.

it to "at least 3° in 300^y", and then attributed the latter value without the words "at least" to Hipparchus. There are 266^y between the epochs of Hipparchus' and Ptolemy's catalogs. The cumulated precession in 266^y at 1°/cy is 2° 40'; the cumulated precession at the correct rate is a bit larger than 3° 40'. The difference is close to the longitude error in the star catalogs.

If the error in solar position arising from the bad calculation had been transferred to the stars, the precession rate would have seemed to be about 1°/cy, and Ptolemy could have thought that this value and Hipparchus' value were equal within the accuracy of the latter.

There is probably no way to decide how Ptolemy constructed the star catalog. I incline toward the position, as did Peters and Knobel [1915] and others, that he (or perhaps the trusted colleague) adapted Hipparchus' catalog by adding 2 2/3° to Hipparchus' longitudes. If we were dealing with experimental errors, the argument that accounts for all errors by a single basic error (in the longitude of the sun) would be powerful. Here the basic error is the substitution of calculated for observed values; an argument which says that this was done only once rather than twice is not strong. Once it is done at all, repetition is easy.

However, Ptolemy probably did make some independent observations. He "confirmed" Hipparchus' value for the length of the month, but he used observations from his own time to correct the rates of the lunar node and perigee and to add to Hipparchus' theory the lunar perturbation that we now call the evection [Pannekoek, 1961, p. 153]. Further, he gave some stellar declinations separate from the star catalog [Pannekoek, 1961, p. 150]. A precession rate of the equinox can be calculated by comparing these declinations with those given by Hipparchus about 266 yr before and with those given by Timocharis and Aristyllus about 170 yr before Hipparchus. The result is 46"/yr. The average of the correct rate of 50"/yr for 170 yr and of 36"/yr for 266 yr is only 41$\frac{1}{2}$"/yr. Thus Ptolemy's declinations may represent direct observations.

In summary, the equinox and solstice times given by Ptolemy were probably calculated and not observed. [†] Whatever the way by which they were obtained, they are useless for estimating the acceleration of the earth's spin.

Have Ptolemy's errors left any residuum in the length of the year or in other astronomical constants?

If Ptolemy "reconciled" his and Hipparchus' equinox observations, he could have done so by altering the times of either set. However, the results shown in Table II.1 indicate strongly that he did not alter the data received from Hipparchus. Not only do Hipparchus' values agree well with those calculated from modern tables while Ptolemy's disagree violently, but Hipparchus' values show "discordances" with what Ptolemy believed. We have reason to believe that at least one of the discordances is real. Thus it seems probable that Ptolemy transmitted older data, both these and data to be used later, without alteration.

TABLE II. 3

ISLAMIC EQUINOX AND SOLSTICE OBSERVATIONS

Date	Place	True Local Time	Longitude (deg)	JSD - 2 000 000	Δt^c (hr)	$10^9(\dot{\omega}_e/\omega_e)$ cy^{-1}
830 Sep 19	Damascus	$12^h.13$	36.3	24 476.899	0.22	4.4
830 Sep 19	Bagdad	13^h	44.4	24 476.913	0.56	11.2
831 Mar 17	Bagdad	02^h	44.4	24 655.465	-2.21	-44.1
831 Sep 19	Bagdad	19^h	44.4	24 842.163	0.82	16.4
832 Mar 16[a]	Bagdad	08^h	44.4	25 020.715	-1.85	-37.0
832 Jun 18	Bagdad	00^h	44.4	25 114.377	0.98	19.6
832 Sep 18	Damascus	$23^h.3$	36.3	25 207.365	-0.17	- 3.4
844 Sep 18	Bagdad	$21^h.4$	44.4	29 590.263	-0.48	- 9.8
851 Sep 19[b]	Nisabour	12^h	58.8	32 146.831	-3.50	-72.7
882 Sep 19	ar-Raqqah	01^h15^m	39.0	43 469.438	-1.10	-24.3

[a] 17 according to the translator of Ebn-Iounis   [c] Δt = solar time - ephemeris time

[b] 18 according to the translator

[†] Since writing this section, I have discovered that Britton [1967], in an unpublished dissertation, has made the calculations relating to the equinoxes but not to the solstice. I am indebted to Dr. Owen Gingerich of the Smithsonian Astrophysical Observatory for calling my attention to Britton's dissertation. Britton also concluded that the equinoxes were calculated.

3. ISLAMIC EQUINOX AND SOLSTICE OBSERVATIONS

Ebn Iounis [1008] recorded a number of equinox observations and one solstice observation[†] made a century or more before his time. The dates, places of observation, and local times of observation are listed in the first three columns of Table II. 3. The observation of 882 Sep 19 is a famous one made by Al-Battani; the year listed[‡] is the one given by Nallino in the translation cited.

Approximate coordinates of the places listed in Table II. 3 are: Damascus, 33°.5N, 36°.3E; Bagdad, 33°.3N, 44°.4E; Nisabour[*], 36°.2N, 58°.8E; and ar-Raqqah[*], 36°.0N, 39°.0E.

The timing error expected in Islamic observations will be discussed in Section V. 4. The expected error is

---

[†] Someone has already used these observations. In the copy of the work owned by the U. S. Naval Observatory, there are pencilled notes connected with each observation. In the margin opposite the Arabic text, the dates and times are pencilled in. The pencilled notes are clearly taken from the Arabic text and not from the French translation, since they differ from the French in a few cases. In the margin opposite the French text, the days and times calculated from tables are pencilled in, with occasional comments in French. This person was apparently fluent in modern French and in the Arabic of the time of Ebn Iounis. His reading of the dates proves to be correct in all cases but one (see the next footnote), as calculation shows. I should appreciate hearing from any reader who can identify this person.

[‡] Pannekoek [1961] gives the year of Al-Battani's observation as 880. The cited translation of Ebn Iounis gives the year as 883; since the unknown commentator (see the preceding footnote) also gives this year, the mistake is probably in the Arabic text. Al-Battani used this observation and one of Ptolemy's to find the length of the year. Use of the year as 882 gives Al-Battani's results.

[*] On the assumption that Nisabour is the present Nishapur or Neyshabur and that ar-Raqqah is the same as Raqqa.

negligible for the needs of this section. There are two problems in interpreting the kind of time being used.

Two kinds of hour were used in ancient times. They have been known by various names; in this work they will be called equal hours and unequal hours. The equal hour is the newer kind, at least in western usage. It is 1/24 of a solar day. An unequal hour of the night was 1/12 of the interval from sunset to sunrise and an unequal hour of the day (that is, of daylight) was 1/12 of the interval from sunrise to sunset. The unequal hour vanished from use when accurate time keeping became commonplace. At the equinoxes all hours are equal, and there is no problem of interpretation. At a solstice the difference between equal and unequal hours is greatest. However, for the one solstice given in Table II. 3, the time is midnight, which is the same whether expressed in equal or unequal hours. Hence there is no problem for this observation either. However, there will be a problem for some of the time measurements to be used in Section V. 4.

The other interpretation problem concerns the difference between mean and apparent time. The general size of the equation of time (defined as mean minus apparent time) is about the same as the expected timing error. However, the question of mean versus apparent time involves a possible bias and is more important than a clock error.

In the original form of the record, the time is usually given as a certain number of hours from one of the four main time points of the day, that is, from sunrise, noon, sunset, or midnight. This suggests but does not prove that the times are apparent. One would expect the adoption of mean time to be accompanied by a standardization of the reference point. Hence I shall assume that the times listed in the third column of Table II. 3 are apparent local time.

The equation of time will not be calculated for each observation. It will be assumed that the equation of time equals $+8^m$ at each vernal equinox, 0^m at the summer solstice, and -8^m at each autumnal equinox. The errors in these values probably do not exceed 1^m.

The longitudes used in converting from local to Greenwich time are listed in the fourth column of Table II. 3. The values of JSD calculated on the basis of the assumptions that have been described are listed in the fifth column.

The values of JED were obtained by inverse interpolation in the tables of Nav. Obs. [1953]. The values of JED are not needed directly and are omitted from Table II. 3 for brevity. JED can be found from Δt, which is listed in the sixth column. Finally, the value of $10^9(\dot{\omega}_e/\omega_e)$ deduced from each observation is listed in the seventh column.

As one would expect, there was improvement in accuracy between Hipparchus' observations in Table II. 1 and those in Table II. 3. The values of Δt in Table II. 3 have not been corrected for bias in the setting of the equatorial circles, as were Hipparchus' values. However, they differ from the correct value, which should be about -1^h, by only about 2^h or less, indicating that the equatorial circles were set within $2'$ or better. The error in Hipparchus' equator was estimated at about $4'.5$.

Since several observatories are involved, there is no single equatorial bias. If the biases were random, however, it is legitimate to use all the values in Table II. 3 without correction for bias. Since the precision in finding a solstice is poorer than that in finding an equinox, at least if the solstice were found by the direct method of finding when the declination of the sun was a maximum, the weight attached to the solstice observation should be lowered. Rather arbitrarily, the solstice observation will be given a weight of $\frac{1}{4}$ compared with 1 for each equinox.

The weighted mean and the residuals of the values of $10^9(\dot{\omega}_e/\omega_e)$ in Table II. 3 give the estimate

$$10^9(\dot{\omega}_e/\omega_e) = -17 \pm 9 \text{ cy}^{-1}. \qquad (II. 7)$$

27

The first four equinox times measured at Bagdad certainly form part of a set. They could be combined to give an estimate of an equator bias and of Δt. However, they are in successive years and are given only to the nearest hour. They are in fact consistent with the length of the Julian year and pose the same problem of interpretation that we had with Hipparchus' values. It would be necessary to have values over a number of years, if they are reported only to the hour, in order to obtain a significant result.

The fifth equinox at Bagdad, in 844, may or may not be part of the series. It is reported to a precision[†] of better than an hour, so the observer probably changed, but there is a good chance that the instrument was the same. The succession $0^h.56$, $0^h.82$, and $-0^h.48$ of values of Δt for the autumnal equinox times at Bagdad go in the way that we would expect. Assuming that all five Bagdad equinoxes form a set, we take the mean of Δt for the vernal equinoxes; this is $-2^h.03$. We take the mean of Δt for the autumnal equinoxes; this is $+0^h.30$. The mean of these is $-0^h.86$. The bias indicated for the equator setting is slightly more than $1'$. There is no strong indication of the error to assign to the mean value of Δt. It is plausible to assign the standard deviation that would result from rounding times to the hour; this is close to $0^h.3$ standard deviation.

Thus the set of equinox readings from Bagdad yields the estimate $\Delta t = -0^h.86 \pm 0^h.30$. The mean value of T is about -10.6 cy. The result is

$$10^9(\dot{\omega}_e/\omega_e) = -17 \pm 6 \text{ cy}^{-1}. \qquad (\text{II. 8})$$

Thus the estimate formed from the Bagdad readings alone is the same as that found in Eq. (II. 7) from the set of all values, except for being assigned a smaller standard deviation. The estimate of the standard deviation in (II. 7) was

---

[†] The time is given as $23' 25''$ of the day. $(23/60) + (25/3600) = 0.3903$; this fraction of a day equals $9^h 22^m$, approximately. The intended precision is unknown, but may have been $5''$ of a day, or 2^m. The true precision is certainly not this good.

certainly high since it absorbed the bias in the equator. I shall therefore assume that Eq. (II. 8) gives the best estimate to be found from Table II. 3.

If the solstice in Table II. 3 is weighted equally with the equinoxes, the central value of $10^9(\dot{\omega}_e/\omega_e)$ is changed to -14 cy^{-1}. If the value from the year 844 at Bagdad is omitted from the Bagdad set, the central value in Eq. (II. 8) is changed to -13 cy^{-1}. Thus, the reader is justified in using -14 ± 6, say, in place of the value in Eq. (II. 8) if he wishes.

4. MISCELLANEOUS OBSERVATIONS OF SOLAR AND LUNAR POSITION

Ptolemy [ca 152; III. 2] transmitted an observation of the summer solstice made by "pupils of Meton and Euctemon" in Athens in -431. Since Hipparchus felt that the accuracy of his measurements of equinoxes, made about three centuries later, was hardly better than $\frac{1}{4}$ day, and since the precision of a solstice observation is much poorer than that of an equinox[†], it is to be expected that the precision of this measurement is of the order of a day. This observation will hardly be useful; it is included as a curiosity, being the oldest known record of solar position.

Gaubil [1732a; pp. 71, 72, 93, 97, and 102] gave the times of five observations of the winter solstice made in China in years ranging from 618 to 1198. There is no a priori estimate of the accuracy of these observations; however, it is reasonable to use an a posteriori estimate based upon the scatter of the individual values. The readings were made at different times and places, but they can be combined safely because there is no bias term in a solstice observation as there is in an equinox. The weight

---

[†] Because a solstice requires finding the time of an extremum of solar declination while an equinox requires the time of a zero.

assigned to an individual reading tends to decrease as the date approaches the present. This tendency is countered by the effect of improved observational techniques. Since no better choice is available, the Chinese solstices will be assigned equal weights. The calculated results will indicate in fact that this assignment is reasonable.

Athens has coordinates 38°. 0N, 23°. 7E. "Siganfou" is Ch'ang-an, the capital of the Former Han dynasty, and the modern Sian, in Shensi province; it is at 34°. 2N, 109°. 0E. Gaubil [1732a] does not directly identify Teng-fong, although he does give (page 139) the latitudes and the longitudes (the latter measured from Peking) of a number of Chinese cities. However, on page 77 he says that Teng-fong is the same as Yang-tching in Honan province. He also adds the information that a gnomon 8 feet high casts a shadow 1 foot 4 inches long at Yang-tching, which indicates a latitude near 34°N. Thus Teng-fong is probably Yenching at 33°. 6N, 114°. 0E. Caifong-fu is Kaifeng at 34°. 8N, 114°. 4E, and Hang-tcheou is Hangchou at 30°. 3N, 120°. 2E. Peking is at 39°. 9N, 116°. 4E. Except for Teng-fong, the places can be identified not only by pronunciation but by Gaubil's table of positions.

TABLE II. 4

MISCELLANEOUS SOLSTICE OBSERVATIONS

Date	Place	True Local Time	Longitude (deg)	JSD	Δt^a (hr)	$10^9 (\dot{\omega}_e / \omega_e)$ cy^{-1}
- 431 Jun 27	Athens	06^h	23.7	1 563 812.684	-32.8	-138
618 Dec 19	Siganfou	$15^h.4$	109.0	1 947 134.838	8.4	118
727 Dec 18	Teng-fong	$17^h.3$	114.0	1 986 945.904	- 1.4	- 23
980 Dec 16	Caifong-fu	$15^h.1$	114.4	2 079 352.813	9.7	263
1179 Dec 15	Peking	$03^h.4$	116.4	2 152 035.317	- 9.9	-437
1198 Dec 15	Hang-tcheou	$04^h.1$	120.2	2 158 975.338	- 0.1	- 6

[a] Δt = solar time - ephemeris time

Information concerning the Chinese solstices and the solstice of Meton and Euctemon is summarized in Table II. 4, which has the same format as Table II. 3.

The value of Δt for the solstice of Meton and Euctemon is $-32^h. 8$. Biot [1849] used the date given by Ptolemy for this solstice as an illustrative example in converting dates, and remarked that the date found is a day early. He also said that there is no way to check the calculation of the date, which is given in terms of "Metonic cycles", because there is no date given in terms of Metonic cycles that can be found independently. However, the reduction given agrees with what Ptolemy would have given had he used the Julian calendar. He was close enough to the time that he probably had ways of determining how to use the Metonic cycles. Further, since the correct value of Δt is probably around -4^h or -5^h, the error of observation is around, say, 28^h, not far from the a priori estimate. Thus it is likely that the date in Table II. 4 is the correct expression of the measured time when expressed in the Julian calendar.

Because of the low weight that would correspond to its estimated accuracy, the solstice of Meton and Euctemon will be omitted from the final inference of the accelerations.

The estimate of $10^9(\dot{\omega}_e/\omega_e)$ formed from the Chinese solstices is -17 ± 119. Because of the low weights that they would receive, these observations will also be omitted from the final inference of the accelerations.

We now turn to a pair of observations of a quite different character. Ebn Iounis [1008; p. 222 in Caussin's translation][†] gave the mean longitudes of the sun and of the moon for the epoch 1000 Nov 30, at noon Cairo mean solar time. The values given are

---

[†] Strictly speaking, the results about to be quoted are not in the translation but in a summary written by Caussin of some untranslated parts of Ebn Iounis' work. There must be at least two editions of Caussin's translation. Prof. van der Waerden says that the page number is 238 in his copy.

$$L_S = 8^S \ 14° \ 45'57''06''',$$

$$L_M = 9^S \ 00° \ 41' \ 12'' \ 25'''.$$

(II. 9)

The unit denoted by the superscript s equals 30°; perhaps the symbol refers to the angular extent of a signe of the Zodiac. If the longitude of Cairo is 31°. 2E, the epoch is

$$JSD = 2\ 086\ 641.\ 9133.$$

(II. 10)

The values of L_S and L_M were found by combining many observations.

The nutation was unknown at this time, although there had been discussion for several centuries about the possibility of a "trepidation" of the equinoxes [Pannekoek, 1961, pp. 166 et seq.], stimulated partly by the inconsistency of Ptolemy's value of the precession with other values. Data in 1000 were not good enough to confirm or deny such a trepidation. Thus, if the data used to find the values in Eq. (II. 9) spanned one or more cycles of the long-period terms in the nutation, it can be assumed that the values of L_S and L_M are referred to the mean equinox of date, as implied by the notation used in Eq. (II. 9).

The astronomers' ignorance of nutation allows us to guess at the accuracy of the numbers in Eq. (II. 9). No matter how many data went into finding the smoothed values of L_S and L_M, it is unlikely that they are much more accurate than the amplitude of the nutation in longitude. This suggests $20''$ as one limit to the accuracy of Eq. (II. 9).

On the other hand, Table II. 3 indicates that the accuracy of placing an equatorial plane is of the order of $1'$ and that the accuracy of locating the sun in declination (with respect to the biased equator) is also of the order of $1'$. This suggests $1'$ as another limit to the accuracy of Eq. (II. 9).

I shall use $30''$ as the a priori estimate of the standard deviation of L_S and L_M in Eq. (II. 9), or 0°. 0083.

The value of ephemeris time when L_S had the value given in Eq. (II.9) is readily found by using the equation for L_S on page 98 of <u>ES.</u> The result is JED = 2 086 641.9445. In conjunction with Eq. (II.10), this yields

$$10^9(\dot{\omega}_e/\omega_e) = -25.1 \pm 5.6 \text{ cy}^{-1}. \qquad (II.11)$$

Equations (II.9) and (II.10) also give L_M for a particular value of GMST. GMST is related to ET by Eq. (II.4); the relation involves the unknown constant $\dot{\omega}_e$. L_M is related to ET by the equation for L_M on page 107 of <u>ES</u>, on the assumption that $\dot{n}_m = -22''.44/\text{cy}^2$. If $\dot{n}_m$ has a $\overline{\text{different}}$ value, the term $\frac{1}{2}(\dot{n}_m + 22''.44)T^2$ should be added to the equation for L_M.

In order to use the value of L_M from Eq. (II.9), it is necessary to eliminate ET between the equations for GMST and the modified equation for L_M. Rigorous elimination of ET is difficult because of nonlinear terms. Since the nonlinear terms are small, it suffices to evaluate them by use of the specific value of ET already found from the value of L_S.

Elimination of ET and subsequent use of the values of L_M and of GMST is tedious but straightforward, and it does not seem worthwhile to give the details. The result is a linear relation between $\dot{\omega}_e$ and $\dot{n}_m$:

$$\dot{n}_M - 1.7373 \times 10^9(\dot{\omega}_e/\omega_e) = -2.31 \pm 0.74 \text{ }''/\text{cy}^2 \qquad (II.12)$$

if $10^9(\dot{\omega}_e/\omega_e)$ is in units of cy^{-1}. Light time can be neglected for the moon.

To the accuracy needed, this is the same result that we would get by interpreting time in the equation for L_M as GMST rather than as ET, and by interpreting the difference between the calculated L_M and the value from Eq. (II.9) as the quantity called L in Chapter I (see Eq. (I.1)).

We can only hope that there has been no copying error in the transmission of the values in Eq. (II.9). There is no way to check.

Any solar and lunar tables or calculated ephemerides presumably represent smoothed observations. They may be used in a similar way provided certain conditions are met: (a) we must know approximately the mean epoch of the observations that were used in establishing the basic constants of the tables, and (b) the tables must cover the mean epoch or else we must be able to extend them with confidence to the mean epoch. The values given by Ebn Iounis were referred to an epoch in his own time, and his work shows throughout that he was concerned with the discrepancies between current observations and earlier tables. Hence it is probable that the observations back of Eq. (II.9) were close in time to the epoch used and that both conditions are met.

Section II.2 shows that we cannot use Ptolemy's tables with confidence. All other tables from the ancient Greek culture have apparently been lost. There are many cuneiform ephemerides but, so far as I know, we cannot determine the epochs of the underlying observations with confidence. There are probably Islamic tables that I have not studied that can be used with confidence.

5. SUMMARY OF THE RESULTS OF CHAPTER II

The data discussed in this chapter have yielded four equations of condition for $\dot{\omega}_e$ and $\dot{n}_M$, namely Eqs. (II.6), (II.8), (II.11), and (II.12). The combination of these equations yields

$$10^9 (\dot{\omega}_e / \omega_e) = -22.0 \pm 3.2 \ \text{cy}^{-1} ,$$

$$\dot{n}_M - 1.7373 \times 10^9 (\dot{\omega}_e / \omega_e) = -2.31 \pm 0.74 \ ''/\text{cy}^2 ,$$

(II.13)

whence $\dot{n}_M = -40.5 \pm 5.6 \ ''/\text{cy}^2$. These are not necessarily the best estimates to be formed from all the ancient observations. They are merely the best estimates to be formed from a small subset.

CHAPTER III

REMARKS ON SOLAR ECLIPSES AND ANCIENT ECLIPSE RECORDS

1. GENERAL COMMENTS ABOUT TOTAL SOLAR ECLIPSES AND RECORDS

The total eclipse of the sun is a rare and spectacular event. Its dramatic impact has had direct effects upon human conduct. Because it was often regarded as an omen, its indirect effects are incalculable. Whether omen or not, its occurrence may be noted and recorded by almost any writer, not only by a professional astronomer.

It is rare, of course, because the zone of totality is narrow, typically of the order of 100 km. The duration of totality at a particular point is of the order of 4^m. Thus a simple observation of totality, coupled with an observation of a place where totality occurred, is an astronomical observation of considerable accuracy. The time is needed only to identify the eclipse; an accuracy of a decade in reporting the time is enough in some cases. Unfortunately these simple ingredients of an accurate report are often missing.

Many eclipse reports are found in national annals or chronicles that reported events of interest anywhere in the country. Sometimes the exact place can be recovered from the annals, often it cannot. If an eclipse can be assigned only to a region, I shall replace the region by a rectangle on a Mercator projection and calculate the eclipse for the center point. I shall assign standard deviations in latitude and longitude that are equal to the corresponding dimensions of the rectangle divided by $2/3$; this is roughly equivalent to assigning equal probabiliities to each point within the rectangle and zero probability outside the rectangle.

Some reports by an individual do not give the place, presumably because the place did not seem important to him. In usable cases of this sort, a region can sometimes be assigned on the basis of the known habits of the writer.

In some cases the observer can be placed on a line such as a river. In still others he can be placed at one of two points; I shall take this as equivalent to placing him on a straight line joining the two points. When the uncertainty in place narrows from a region to a line, the errors in latitude and longitude are no longer independent. The direction of the path of totality must be considered before assigning a standard deviation to the observation of place. The procedure will be to project from the extremes by lines parallel to the path of totality onto the parallel of latitude passing through the centroid for which the eclipse will be calculated. The total resulting extent in longitude, divided by 2/3, will be assigned as the standard deviation in longitude; the standard deviation in latitude can then be taken as zero.

In considering the accuracy of an observation of totality, one should note first that many reports simply state that an eclipse occurred, with no accompanying detail. When such a report was made by a non-astronomer, I shall usually assume that the eclipse was large enough to be noticed by an ordinary person who was not watching for it. I have only two pieces of data on the necessary magnitude[†].

Hayden [1968] once observed an eclipse that attained a magnitude of about 0.8 where he was. After the eclipse he asked a number of people what they thought about it. No one whom he questioned was aware that an eclipse had happened.

The largest eclipse magnitude that has occurred where I have been present was about $\frac{3}{4}$. I was aware of the eclipse prediction but did not observe it because of the weather. However, judging from the extent of darkening, I do

---

[†]The magnitude of an eclipse will be defined in Chapter V. Here it is sufficient to note that a magnitude of zero means no eclipse and that a magnitude of unity means a central eclipse.

not believe that I would have observed it, even in the absence of clouds, without being alerted to it.

On the basis of these scanty data, I shall assign 0.06 as the standard deviation of the eclipse magnitude when there is no accompanying detail bearing on the magnitude. This is roughly equivalent to a high probability of observation when the magnitude is greater than 0.9 and a low probability when it is less than 0.9. A standard deviation this large makes an individual report with no detail carry little weight. However, there are enough such reports that the collection of them may make a significant contribution.

A detail accompanying several reports is the statement that stars became visible. Neither Hayden [1968] nor Schroader [1968], who have kindly searched at my request, nor I have found much information about the visibility of stars during an eclipse that does not reach totality. Fotheringham [1920] states that there should be no difficulty in seeing Venus when the magnitude is only 7/8. He also says: "In the case of the eclipse of 1912 Apr 17, I have found no record of any star but Venus being seen without instrumental aid except where the eclipse was total, although there were numerous observers along the long zone where the eclipse passed through France, Belgium, and Germany just missing totality." On this basis, he questioned the accuracy of a statement by Thucydides (see the record -430 Aug 3a M in Section IV.5).

In the next year Fotheringham [1921] reversed himself: During the eclipse of 1921 Apr 8, which attained a magnitude of slightly less than 0.9 at Oxford, numerous observers at Oxford and nearby places reported seeing from 2 to 4 stars (including Venus) without optical aids and without foreknowledge of the stellar positions. Then, in a supplementary note, Fotheringham stated that he had learned that the champion in this friendly competition had in fact computed the positions in advance and had used field glasses to see one of the stars (Vega). The gentleman in question still maintained that he saw the other stars with the naked eye. One fears that Fotheringham was the victim of a learned practical joke played by some partisans of Thucydides.

In an attempt to acquire some quantitative data, I watched for the appearance of stars around twilight on 1968 Apr 16 and again on Apr 28. On both occasions the air was exceptionally clear and the sky was free of clouds. I could see one star when the disk of the sun was almost bisected by the visible horizon, as estimated by another observer. I saw a second star when the sun should have been about 2° below the horizon. On the first occasion I saw no additional stars within 20^m after sunset; on the other occasion I had seen five by that time.

Dandekar [1968] finds that the brightness of the zenith sky, for an eclipse that occurred at about 09^h local time, at 99 percent obscuration is about the same as that of the zenith sky when the sun is 2° below the horizon. This is about the condition under which I could see two stars on both occasions. He finds that the sky brightness at mid-totality was about that with the sun 4° below the horizon. This is about the circumstance under which I saw two and five stars, respectively.

It is clear that the results may vary widely for different circumstances. Knowledge of star locations would aid in seeing them. I made no attempt to find out where to look in advance and made no attempt to identify the stars later. I did verify that Venus was below the horizon on both occasions. I made no attempt to achieve dark adaptation except to refrain from looking toward the sun for about $\frac{1}{2}$ hour in advance. It is certain that my eyes were not as well dark-adapted as those of, for example, an inhabitant of a medieval monastic cell. The presence of Venus might affect whether stars were seen.

For want of a better value, I shall assign 0.01 as the standard deviation from totality when star visibility is the only detail accompanying an eclipse report.

References to the corona or to prominences would be useful in judging totality. Curiously, there are few ancient reports into which one can read a description of corona or prominences, and the interpretation of at least two of these is doubtful.

38

In interpreting the ancient records and in assigning standard deviations to them, it must be remembered that many years and sometimes many writers are interposed between the observation and the oldest record that we now have. Whether multiple writers or merely time with the same writer is interposed, I believe that the result is a tendency toward loss of detail and less accuracy. The memory inevitably loses detail with the passage of time. If one writer paraphrases another and drops detail whose importance he does not see, it is impossible to restore the detail unless an earlier text with the detail is available. And, even without fraudulent intent, a person has a strong tendency to increase the dramatic impact of past events.

Because of this problem, because of the terms used, and because of the vagueness of some of the records, use of the eclipse records necessarily involves textual criticism, a task for which I am not well prepared. However, there is no standard interpretation that can be used for many of the eclipses because there is frequent disagreement among the authorities who have studied the texts. Therefore I have had to carry out independent textual criticism for this study.

The result of my textual criticism is summarized by the number assigned to each record, which I shall call the "reliability". Two factors go into this number: first is my assessment of the probability that the information given in a record can genuinely be attached to a solar eclipse; second, in those cases where a unique identification of the eclipse is not possible, I distribute the reliability among the possible eclipses according to an estimate of the probability that the record applies to each.

In using the reliability, the weight to be assigned to each eclipse record is first calculated on the basis of the standard deviations assigned to position and magnitude. This weight is then multiplied by the reliability to obtain the final weight.

Assigning reliability is a dangerously subjective process. Knowledge of a numerical result from a record, in the presence of a preconceived idea of approximately what the numerical result should be, could easily influence the assignment of reliability. Therefore I have studied the texts and assigned the numerical values of reliability before doing any computations with the eclipses, except for the necessary process of looking into Oppolzer [1887] or Ginzel [1899] in order to obtain an identification. Except for editorial changes, and for some trivial substantive changes that will be noted, Part I, including the discussion of the eclipse records, stands exactly as it did before any eclipse computations were carried out.

If two or more identifications are assigned to one eclipse record, this means that two or more possible eclipses happened within the range of time allowed by the record. The identifications to be given in Chapter IV have been chosen by looking on the charts in Oppolzer's or Ginzel's Canons for eclipses within the allowed time range that would have been large in the allowed region. Preparation of the charts involved approximations, and the actual eclipse path may have differed by a considerable amount from the path shown on the charts for many eclipses; thus, some of the tentative identifications to be made in Chapter IV may turn out to be unreasonable on the basis of the detailed calculations to be used in Part II of this work. An identification will be considered to be unreasonable if it turns out to require a value of $\dot{n}_M$ outside the range (0, -45) $''/cy^2$ or a value of $10^9(\dot{\omega}_e/\omega_e)$ outside the range (0, -35) cy^{-1}. An unreasonable identification will be dropped from the final inference, and its share of the reliability will be redistributed among the remaining reasonable identifications. To this extent the findings of Part I may be modified by the calculations to be made in Part II.

A converse process may also occur sometimes. That is, a reasonable identification may be overlooked because of the approximate charted paths or because of simple oversight. I do not see a feasible way to avoid this possibility and can only hope that overlooked identifications will not bias the final results.

40

It is not to be expected that the interpretations and the assignments of weights and reliability in Chapter IV will prove to be accurate or universally accepted in all cases. The hope is that the errors introduced will be random in nature and that the finally inferred accelerations will be accurate, in spite of the individual errors, because of the large number of records used.

Aside from the large number of eclipses studied, the main difference between this study and earlier ones is the statistical use of the reliability and of the standard deviations in position and magnitude. The tendency in earlier work has been to try to choose only eclipse records that the worker believed to be completely reliable and that indicated complete totality for an identifiable eclipse in a known and small geographical area. The number of ancient records that meet all these conditions is quite small, and the choice of records has proved to be highly variable. This accounts somewhat for the wide spread of earlier results.

Chapter IV will present ancient reports of large solar eclipses, divided into sections according to provenance. The discussion of each record will be preceded by one or more references to the location of the record or to earlier discussions of it.

When I cite only a secondary reference it is because the given reference has adequate quotations and cites adequate primary references and because the problems posed by the text seem so small that the secondary reference is adequate.

In much earlier work it has been the custom to identify an eclipse record by the name of an associated person or place. The number of records used here is too large for this means of identification to be feasible. Here, an eclipse record, whether of a solar or a lunar eclipse, will be identified by the calendar date of the eclipse (in GMST) followed by an abbreviation indicating the provenance. The abbreviations and the corresponding areas are:

B: British Isles
BA: Babylonia and Assyria
C: China
E: Continental Europe, excluding Greece and
      European Turkey
Is: Islamic countries
M: Mediterranean countries, including Greece, and
      including pre-Islamic records from areas that
      later became Islamic

For a few eclipses there are independent records from the same provenance. These will be distinguished by a lower case letter following the day of the month. Independent records of the same eclipse will be used in the final inference of $\dot{n}_M$ and $\dot{\omega}_e$ as if they referred to different eclipses.

If the name of a person or place is traditionally associated with an eclipse, I shall put the name in parentheses after the identification.

A table will accompany each section. The table will give the identification and the estimated hour of local time if available from the record. It will give the latitude and longitude of the central location assigned to the eclipse report (this does not mean a place where the eclipse was central in the astronomical sense). Finally, the table will give the standard deviations assigned to the latitude, longitude, and magnitude, followed by the reliability.

The hour, if given, has negligible value in the analysis of the eclipse. It is given in the tables only to assist in judging the surrounding circumstances. It is given in the 24-hour notation. In assigning the hour, I equate sunrise to 06^h and sunset to 18^h if the hour is given with reference to those events. Greater accuracy in assigning time by trying to calculate the times of sunrise or sunset is not warranted since the times will not be used in the analysis. The term "midday", when it is used in the source text, is not taken as an indication of the hour.

Calculations relating to an eclipse record discussed in this work will be carried out sometimes even when the reliability is assigned the value zero. This will help a reader who wishes to make independent inferences on the basis of his own assignment of weights.

2. CLASSES OF ECLIPSE REPORTS AND THEIR RELIABILITY; OR, HOW TO BELIEVE SIX IMPOSSIBLE THINGS BEFORE BREAKFAST

Many eclipse reports can be classified by certain common features. It is possible to make some general remarks about the reliability of each class. It will still be necessary to study each report on its own merits to see whether it should receive the reliability characteristic of its class.

a. Technical reports. Many of the later records were made by technical experts. In spite of the example of Section II.2, one can accept these unless there is strong contradictory evidence. One must remain alert to the possibility of typographical errors.

b. Annals and chronicles. If an eclipse is reported in annals or chronicles, and if no attempt is made to dramatize the eclipse, one can accept the record[†] in the absence of contradictory evidence. Attention must be paid to possible typographical errors in the record or its transmission, and a reasonable amount of confusion in the dating must be forgiven. Dubs [1938; Appendix to Chap. III] gives an example of an annals record that was probably fabricated.

c. The assimilated eclipse. A number of eclipse records were written down some time after the event. Sometimes the writer was probably the observer, sometimes he had only talked to an observer. If another dramatic or important event happened close in time to the

---

[†]If the events were recorded contemporaneously. Annals compiled from older sources long after the event must be used carefully.

eclipse, there is a strong tendency for an observer to assimilate the eclipse and the other event in time. Assimilation may account for some apparent errors in dating
eclipses. If it does not interfere with identifying the
eclipse, and if we can be sure that it is all that has happened,
assimilation probably does not affect the reliability. However, as the time between the event and the record increases,
the risk that the eclipse has been transferred in both time
and place increases, and the risk that the eclipse has been
invented to fit the occasion increases.

d. The magical eclipse. There are at least two
types of magical eclipses.

Since an eclipse was often regarded as an omen, an
imaginative writer could place an eclipse before or at the
beginning of a great event, such as a military campaign,
and interpret it to suit the course of history. This procedure
perhaps accounts for the remarkable tendency of people to
fight battles during a solar eclipse.

The death of a king or an eminent person is often
accompanied by prodigies or marvels. A well known example is given by Plutarch [ca 100; Life of Caesar]: "The
most signal preternatural appearances were the great comet,
which shone very bright for seven nights after Caesar's
death, and then disappeared, and the dimness of the sun,
whose orb continued pale and dull for the whole of that
year ... " Also the fruits did not ripen properly and a
phantom appeared to Brutus.

This kind of magical eclipse perhaps stems from the
once widespread belief that the health and welfare of a
king were coupled by sympathetic magic to the fertility and
welfare of his land and people. His death, or merely his
bad health or blemishment, could lead to the "Waste Land"
[Weston, 1920], and to the disruption of all the ordinary
course of nature [Frazer, 1922].

When an eclipse is surely and purely magical, it
should receive a reliability of zero. However, an apparently

magical eclipse may sometimes have started out as an assimilated eclipse, or as an eclipse used simply as a mnemonic device. An ancient writer with a taste for the marvelous could subsequently have altered such an eclipse, which was once almost factual, into a magical eclipse. We must decide in each case whether a magically used eclipse is assimilated or pure invention.

If a writer merely connects an eclipse to some other event without making a marvel of the coincidence, it is probable that assimilation or, rarely, genuine coincidence has occurred, and the reliability is unaffected by the connection. If the eclipse is made into a marvel, and particularly if the eclipse is one of many marvels, the reliability will be taken as zero. Exceptions will occur in a few cases for which the event can be dated and located accurately by evidence independent of the eclipse, if a large eclipse did occur close in place and time to the event.

e. The literary eclipse. "The rim of black spread slowly into the sun's disk, ... the multitude groaned with horror to feel the cold uncanny night breezes ... and see the stars come out ... (and later) ... the silver rim of the sun pushed itself out ...." [Clemens, 1889]. This is one of the few eclipse reports [†] in which the recovery from the eclipse is specifically mentioned. The omitted parts of the text are not relevant to present purposes.

Clemens, though a classical author, is not an ancient author, but he can be studied by the methods used for one: The quotations clearly show that Clemens saw a total solar eclipse some time before he wrote this work, say between 1850 and 1900 if we pretend that we do not know the exact date of publication. For an ancient work, we usually know the date of writing only within a few decades. We know where Clemens lived and we make a preliminary identification of possible eclipses from the charts in Oppolzer [1887].

[†] Clemens gave the date 528 Jun 21 for the eclipse. Oppolzer's Canon lists four solar eclipses for this year, all partial and none on Jun 21.

45

The eclipses of 1854 May 26, 1869 Aug 7, and 1900 May 28 are good possibilities; others could be admitted with lower probability. We carefully calculate the local circumstances of each eclipse for Clemens' current home, ignoring the possibility that he might have been traveling. We identify the eclipse as the one with the greatest calculated magnitude. We then use this possibility alone in using the eclipse to improve the astronomical constants or the accelerations.

It is doubtful that anyone would analyze Clemens' text in this way, but the pretended analysis illustrates three main points that are relevant to the treatment of some ancient "eclipse records" in the current literature.

First, we can almost surely disprove[†] an eclipse sighting by Clemens on the basis of independent evidence. We cannot usually do this with an ancient writer. Thus ignorance, the inability to disprove, has often been equated with an assessment of high reliability.

Second, if the reader will study carefully the rules of the peculiar game that has been played, which I shall call the "identification game", he will see that they lead to a successful "identification" for almost any set of times and places chosen at random. It is only necessary for there to be a modest uncertainty in either the time or place. Further, the calculated path of the "identified" eclipse, by the rules of the game, passes close to some chosen point. Thus, if the "eclipse" report is used to improve the constants that went into the eclipse computation, by making the calculated path go through the chosen point, it is almost guaranteed that the changes in the constants will be acceptably small[‡].

[†] Or prove. I have not checked the possibility that Clemens saw an eclipse; whether he did or not is irrelevant to the argument.

[‡] In fact, the operational definition of an "identification" makes the "correct identification" equal to the one that minimizes the changes to be made in the pre-assigned constants.

Finally, Clemens did not need to see an eclipse at all in order to write the passages quoted. He was a skillful and imaginative writer and could have written the passages on the basis of his reading alone. Alternately, his use of an eclipse could have been suggested by seeing a small one or by seeing an eclipse prediction in a newspaper or almanac.

By a "literary eclipse", I shall mean one that has been inserted into a work of conscious literary invention. A literary eclipse should be assigned a reliability of zero; it usually comes from imagination and not observation. An exception should be made only if there are strong grounds for believing that the writer did not have access to written or oral descriptions of eclipses[†].

Magical and literary eclipses can be put into a family that can be called myth. It will turn out, if my interpretations are right, that many eclipse reports being used seriously in current literature are myth by any definition. Myth dies hard and is eternally reborn. In the pages that follow we shall see twentieth century scientists begetting new myths. I have tried to avoid generating or propagating myth, but I dare not hope for complete success. The most that I can hope is that I have decreased the amount of myth connected with ancient solar eclipses.

[†] If independent evidence should be found that a writer had in fact witnessed an eclipse, that evidence should be used, but not the literary description.

ANCIENT RECORDS OF LARGE SOLAR ECLIPSES

1. ECLIPSE REPORTS FROM THE BRITISH ISLES

Table IV. 1 is to be used with this section.

TABLE IV. 1

LARGE SOLAR ECLIPSES REPORTED FROM THE BRITISH ISLES

Identification	Hour if Given	Central Location		Standard Deviations			Reliability
		Lat. (deg)	Long. (deg)	Lat. (deg)	Long. (deg)	Magn.	
664 May 1 B	16	53.0	-1.0[a]	1.2	1.2	0.06	1
733 Aug 14 B		51.3	1.1	0.5	0.5	0	1
809 Jul 16 B	11	53.0	-1.0	1.2	1.2	0.06	1
878 Oct 29 B		53.0	-1.0	1.2	1.2	0.03	1
1133 Aug 2a B		51.9	0.4	0	b	0.02	0.5
1133 Aug 2b B	12	51.6	-2.1	0	0	0.01	0.5
1140 Mar 20a B	15	51.9	0.4	0	b	0.02	1
1140 Mar 20b B	15	51.6	-2.1	0	0	0.01	0.5
1191 Jun 23 B	09	51.1	-1.3	0.5	1.0	0.06	1

[a]A negative longitude is measured west from Greenwich.

[b]Cannot be assigned until path direction is calculated. See Section III. 1.

664 May 1 B. Reference: Bede [731; Bk. III, 27].
The translation mentioned in the citation reads: "In the same year of our Lord's incarnation, 664, there happened an eclipse of the sun, on the third of May, about ten o'clock in the morning[†]. " The eclipse is also mentioned, with less detail, in the chronological summary in Book V, 24. This is the last of three solar eclipse reports by Bede. The other two are listed as 538 Feb 15 E and 540 Jun 20 E.

The mistake of two days in the date[‡] does not interfere with the identification. The reliability is unity. The

---

[†]The translator was apparently betrayed by his modern habits. The original has "hora circiter decima diei", or "around the tenth hour of the day". This is about 4:00 PM or 16^h; this time agrees with the chart in Oppolzer's Canon.

[‡]A reference that I forgot to record suggested that Bede's source had the year and month of the eclipse, and that Bede

standard deviation of magnitude will be taken as 0.06. The place needs discussion.

The discussion of the records 538 Feb 15 E and 540 Jun 20 E gives some tentative reasons for placing the source of those records in Italy. One reason is a low probability that records of this sort would have been kept in Britain in 538 and 540. This reason does not apply in 664. We learn from Book II, 14 of Bede that York had a bishop and a wooden church in 627 and that a stone church was begun there in that year. By 664 there were many church establishments in England and records of the type needed in order to account for the entry almost surely existed.

There is the additional fact that the eclipse record for 664 comes in the middle of the main text surrounded by events that are placed in Britain. The reports for 538 and 540 occur only in the chronological summary. Also, in Book IV, 1, Bede, after moving forward in time to 665, comes back to 664 using the words "In the above-mentioned year of the aforesaid eclipse..." He uses this phrase to date an event that happened in England. Because of the way the eclipse occurs in Bede, I do not hesitate to place the observation somewhere in England. For the purposes of this study, England can be taken as the region from 51° to 55°N and from 3°W to 1°E.

733 Aug 14 B. Reference: Whitelock et al. [1961]. Most texts of the Anglo-Saxon Chronicles read something like "In this year Aethelbald occupied Somerton, and there was an eclipse of the sun." Although the month and day are missing, the identification of the eclipse is certain. The text known as F is the only one that adds: "and all the circle of the sun became like a black shield." Oppolzer's Canon shows that this eclipse was annular, and this unique description sounds like an eye-witness account by an observer who was within the zone of annularity. The location of this observer is important.

_____

supplied the day of the month from the tables of the ecclesiastical moon [ES, p. 422], knowing that the eclipse had to occur at the new moon.

The Chronicles were originally compiled [Plummer, 1910] around 890 under Alfred, and probably at or near Winchester [Parker, 1968]. The original Chronicles drew upon Bede [731] and other sources for back material; afterward, they were kept more or less up to date. A copy or copies were subsequently carried north and received both current and old additions there. A copy of a northern version was subsequently taken to Canterbury and there, perhaps in the 11th century, it was combined with local records to make the F text. It is plausible to assume that the unique contribution of F came from a local Canterbury record. I shall take Canterbury[†](51°.3N, 1°.1E) as the place, but shall assign standard deviations of 0°.5 in both latitude and longitude as a precaution against overweighting the assumption.

Standard deviation of magnitude: 0. Reliability: 1.

809 Jul 16 B. Reference: Whitelock et al. [1961]. Most texts read something like: "In this year there was an eclipse of the sun at the beginning of the fifth hour of the day on 16 July, the second day of the week, the 29th day of the moon." F again has a unique contribution; he gives the same English, but in his Latin he makes it "on Sunday, the 12th day of the moon." There was a partial eclipse of the moon on Sunday, 809 Jul 1. From F's contribution we learn the following: (a) the compiler of F in Latin probably mixed up the lunar and solar eclipse records; (b) the weather was good at least twice in 809 Jul; (c) the Latin compiler probably supplied the age of the moon from the tables of the ecclesiastical moon [ES, p. 422]; and (d) the Latin and English versions are not necessarily translations of each other and had some independent sources.

[†] I have recently learned that F copied the clause in question from a northern record called "Continuation of Bede". My error in assigning the place has no astronomical consequences, because the eclipse 733 Aug 14 cannot be used, for a reason unrelated to the record (see Section XIII. 2).

For the solar eclipse I take the following: Place, England; standard deviation of magnitude, 0.06, reliability, 1. Since neither the hour nor the magnitude of the lunar eclipse was recorded, the lunar eclipse report does us no good.

Newcomb [1910] does not include the eclipse of 809 in his list of eclipses visible in England.

878 Oct 29 B. Reference: Whitelock et al. [1961]. Most Chronicles texts read something like: "And the same year there was an eclipse of the sun for one hour of the day." If the entry for this year were the only reference to this eclipse in the Chronicles, one should assign the value 0.06 to the standard deviation of its magnitude. However, in the entry that is dated 885 in most texts but 886 in the C text, it is stated that Louis (II of France) died "in the year of the eclipse of the sun". The entry for 878 makes no attempt to connect the eclipse and the death, so this is not a magical eclipse. Since it was used as a reference date seven years later, it may have been more impressive than the average observed eclipse. I tentatively assign the value 0.03 to the standard deviation of magnitude.

There is an ordinary amount of confusion about what year it was in this section of the Chronicles. The confusion is not enough to cause a problem in identification.

Place: England; reliability: 1.

1133 Aug 2a B. Reference: Whitelock et al. [1961]. Only the E text deals with this and later years. The first part of the entry for 1135 reads: "In this year King Henry went overseas at Lammas, and the next day, when he was lying asleep on board ship, the day grew dark over all lands, and the sun became as if it were a three-nights'-old moon, with stars about it at midday. People were very much astonished and terrified, and said that something important would be bound to come after this - so it did, for that same year the king died..."

King Henry (I of England) died in 1135 and the eclipse was in 1133. This record gives some sort of mixture of the assimilated, the magical, and the literary eclipse. However, there are ameliorating considerations. The chronicler did not try hard to make the eclipse magical since he attached no other marvels to the king's death. He also has the calendar date correct[†]. He did not distort the sequence of events in order to assimilate the eclipse to the year of Henry's death; he had no entries from 1132 until 1137 except that for 1135. I shall use 0.5 for the reliability of this record, although a higher value could be defended.

Only the crescent phase is mentioned, so the eclipse probably did not reach totality, although stars are mentioned. A three-nights'-old moon has 0.08 or more of its surface visible. If this much of the sun were visible, it is doubtful that any star other than Venus could be seen. It is probable that this detail is merely an attempt to emphasize the smallness of the crescent and not a quantitative measure of magnitude. The detail of the stars would normally mean assigning 0.01 as the standard deviation of the magnitude. Because the crescent is explicitly mentioned, I shall increase this to 0.02.

The description of the eclipse is probably local. Plummer [1910] puts E at Peterborough after 1121, apparently on the basis of the dates at which the hand-writing changed. Parker [1968] cautions that it could still have been at Canterbury at this time. I shall assign Canterbury and Peterborough (52°.6N, 0°.2W) as the two possible places for this observation.

1133 Aug 2b B. Reference: William [ca 1143]. The cited reference gives the translation: "For on that very day the sun, at the sixth hour, covered his shining head with gloomy rust, as the poets are wont to say, putting fear into men's minds by his eclipse . . . . In the eclipse I saw myself the stars around the sun . . . ."

[†]Lammas or Lammas Day is Aug 1, and the eclipse is put on the next day.

William identifies the year by saying that Henry completed the thirty-second year of his reign the day before. That would make the year 1132 rather than 1133, but the implied error in reckoning is of a sort that often happens, as any arithmetic teacher can testify. He puts the eclipse and the sailing on the same day, contrary to the preceding record, and gives the day as "nonis Augusti" or "none Augusti" according to the case ending required. William generally used the church (or Roman) calendar; if he meant to do so here, we should read this as the "nones of August", or Aug 5. If he used civil style here, it should of course be Aug 9. William also states that the day was Wednesday. In 1133, Aug 2 and Aug 9 were on Wednesday. It is possible that William used the civil style here and that his eye slipped a week in reading from his calendar if he had one. This explanation makes all the dates associated with the eclipse accord, but I do not push it. In any case, there is no problem about the identification of the eclipse; the precise date is still often confused in records of this era.

Malmesbury (51°.6N, 2°.1W) will be taken as the place.

William assures us that he saw the stars around the sun. He also uses the phrase "tetra ferrugine" that was translated as "gloomy rust". The translator connects the phrase with Virgil's Georgics, I, 1. 467. Ginzel [1899] quotes Virgil's passage, which was written in connection with Caesar's death. Virgil uses "ferrugine" but not "tetra". "Tetra" can mean horrible or hideous as well as gloomy. "Ferrugine" can refer to gray; it can also mean the color of iron rust or the color of the tarnish on bronze. It is hard to say exactly how the phrase should be translated, but it does sound like an attempt to describe the faint light around the sun during a total eclipse. If this description were original with William, it would be reasonable to take the eclipse as total at Malmesbury. Since William was acquainted with the passage from Virgil, I shall use instead a standard deviation of 0.01 for the magnitude.

The reliability is taken as 0.5. See 1140 Mar 20b B for the reason.

1140 Mar 20a B. Reference: <u>Whitelock et al.</u> [1961]. "Thereafter during Lent the sun and the day were eclipsed[†] about noontide of the day, while men ate, so that men lighted candles to eat by; and that was XIII kalends of April[‡]; and men were greatly astonished. "

I shall take the reliability of the record to be unity. The place will be taken as either Peterborough or Canterbury. The detail about lighting candles at the noon meal is unique and reads as if the eclipse has a large magnitude. In the absence of reference to the stars, I shall give a standard deviation of 0.02 to the magnitude.

1140 Mar 20b B. Reference: <u>William</u> [ca 1143], who writes, in the cited translation: "That year in Lent, on March 20, at the ninth hour* on a Wednesday, there was an eclipse, all over England, I have heard. With us certainly and all our neighbours it was such a remarkable eclipse that men sitting at table, as they mostly were at the time, it being Lent, feared the primeval chaos; then, learning what it was, went out and saw the stars around the sun. "

[†]Or grew dark.

[‡]That is, Mar 20.

*This report has "hora nona", the ninth hour or nones. The preceding report has "nontid daeies", noontide of the day. Neither term, when used at this stage of history, can be translated confidently into clock time without external aid. Noon comes from nones or <u>nona,</u> which originally meant the ninth hour of the day or about 3:00 or 4:00 PM depending upon the length of the day. The meaning was gradually transferred to 12:00. The eclipse chart in Oppolzer's <u>Canon</u> indicates strongly that the original meaning of "nones" should be read in both passages.

This report is assigned to Malmesbury. A standard deviation of 0.01 in magnitude will be used because of the reference to the stars. William's feeling that this was a remarkable eclipse, combined with his description of the eclipse of 1133, may indicate that his indications of magnitude are literary, although one may assume with confidence that he saw both eclipses. For this reason, a reliability of 0.5 will be used for both of William's reports.

1191 Jun 23 B. Reference: Appleby [1963]. In Appleby's edition of the Chronicle of Richard of Devizes, page 35, appears: "There was an eclipse of the sun about the third hour of the day. Those who do not understand the causes of things marvelled greatly that, although the sun was not darkened by any clouds, in the middle of the day it shone with less than ordinary brightness. Those who study the working of the world, however, say that certain defects of the sun and moon do not signify anything."

Richard, apparently a native of the town of Devizes, was a monk of St. Swithun's in Winchester, where he wrote his Chronicle. The eclipse record is not dated and it appears in the working margin of the manuscript, a wide space that Richard left for the apparent purpose of putting down material omitted from his first draft. He presumably meant to prepare a clean second draft; if he did, the new manuscript has never been found. His marginal material was just as accurate as the first written material. It is doubtful that the record was noted on the day of the eclipse because it occurs approximately opposite events running to 1191 Jul 28. However, the identification seems secure. Since the account merely says that the sun "shone with less than ordinary brightness", it does not seem likely that the eclipse was total; the conventional value of 0.06 for the standard deviation of the magnitude seems appropriate. The reliability is 1.

There is no indication of the place of observation. The fact that Richard lived at Winchester does not mean that the eclipse was seen there. There is nothing local or personal about the record, as there was about both British records of the eclipse of 1140 Mar 20, to give us confidence that the writer was an eye witness. However, the account does sound as if the writer had some discussion or did some special reading about eclipses, and this suggests that the eclipse was called particularly to Richard's attention. Further, according to Appleby, page 91, the Historical Works of Gervase of Canterbury mention and correctly date the eclipse. Instead of using Gervase as a second source, I shall use his record only to give strength to a location in southern England for the eclipse sighting. I shall take Winchester (51°.1N, 1°.3W) as the central place, with a standard deviation of 0°.5 in latitude and 1°.0 in longitude.

According to Newcomb [1910], the eclipses of 1185 May 1 and 1191 Jun 23 were recorded in the Anglo-Saxon Chronicles. The last entry in any document that is usually called part of the Chronicles is 1154, so perhaps Newcomb intended to write "chronicles", meaning merely early chronicles from England. The report above may be the one Newcomb had in mind for 1191 Jun 23. I have not found a British record of the eclipse of 1185 May 1, but a Norse source is discussed under the heading 1185 May 1 E.

2. ECLIPSE REPORTS FROM BABYLONIA AND ASSYRIA

Table IV.2 is to be used with this section.

TABLE IV. 2

LARGE SOLAR ECLIPSES REPORTED FROM BABYLONIA AND ASSYRIA

Identification	Hour if Given	Central Location		Standard Deviations			Reliability
		Lat. (deg)	Long. (deg)	Lat. (deg)	Long. (deg)	Magn.	
-1062 Jul 31 BA		31.5	45.5	0.9	0.9	0	0
- 762 Jun 15 BA		35.8	43.7	0.9	0.9	0.06	1

Newcomb [1910] said that the eclipse of -1069 Jun 20 was recorded at Babylon, but did not give the source of this information. I have seen no other reference to such a record. The chart in Oppolzer's Canon shows the path of totality passing along the southern coast of the Arabian peninsula, and thence across the ocean and into northern India. It is doubtful that this eclipse was large at Babylon.

-1062 Jul 31 BA (Babylon). Reference: Fothering-Ham [1920]. There is much doubt that the event referred to was a solar eclipse. The passage reads, in the translation given by Fotheringham: "On the twenty-sixth day of the month Sivan in the seventh year the day was turned to night, and fire in the midst of heaven . . ." Fotheringham discusses the passage at length.

If the passage refers to the eclipse of -1062, both the month and the day of the month are given wrong. The passage does not mention the sun directly and it does not use the standard Babylonian term for an eclipse of the sun. There are plausible explanations for all these discrepancies, but the combination of so many of them in one record does not inspire confidence. There is apparently no independent way to date the record except perhaps to the century; no source consulted has mentioned an independent dating.

There are precedents in Babylonian and other literature for using "day turned to night" and heavenly fires in reference to phenomena other than eclipses. Fotheringham cites several of them; the Anglo-Saxon Chronicles for 793 [Whitelock et al., 1961] contain another reference to fires in the heaven. "Fire in the midst of heaven" is a standard Babylonian term for a meteor. However, the description is a reasonable one for a solar eclipse and the failure to use the standard terms would be reasonable if the writer were not a professional astronomer or augur.

Fotheringham [1920] argued for accepting the event as an eclipse on the basis that it is "certainly remarkable that there should have been an eclipse which may, consistently with other evidence, have been total at or near Babylon at or near the date to which the text refers, and that it should have fallen at the right time of year." The coincidence is not remarkable; it is almost necessary. The record means that a certain event (or an alleged event) happened in Babylonia in, say, the -11th century and that it happened within about a month of the same time of year as the eclipse of -1062. Several other large eclipses occurred in Babylonia in that century, at various seasons, and one of them occurred within about a month of the same time of year as almost any random event.

Some papers have cited this as a total eclipse at Babylon on the authority of Fotheringham [1920]. Fotheringham did not conclude this. He gave arguments on both sides of the question and concluded (p. 124), but not firmly, that "the phenomenon recorded in the Babylonian chronical was something other than an eclipse, or, if an eclipse, was total in southern Babylonia and not at Babylon itself . . ." Fotheringham [1958] and Smith [1958] omitted this as an eclipse record in spite of its antiquity.

The text in which the record occurs seems to be a book on portents. A genuine eclipse could of course have been taken as a portent, but there is no reason to take seriously an unverifiable and undatable portentous "eclipse". Because several errors in the writing of the text must be assumed in order to make this portent into a possible eclipse, I shall give this record a reliability of zero, but shall calculate it for the benefit of the reader who may wish to use it as an eclipse.

If the event is taken as a large eclipse that occurred within about a month of the time given, there is no difficulty about identification.

The place may be anywhere in Babylonia. I have assigned the region bounded by latitudes 30° and 33°N and by longitudes 44° and 47°E. If the record is taken to be an eclipse record, "fire" should probably be taken as a reference to the corona or to a prominence, although other possible references to light around the sun do not imply as much brightness as this term does to a modern reader. On this basis the eclipse will be calculated as if it were total.

-762 Jun 15 BA (Eponym canon). References: Fotheringham [1920], [1958] and Ginzel [1899], who give other references. "Insurrection in the city of Assur. In the month Sivan the sun was eclipsed." Independent evidence, due in great part to the lists of kings and their reigns, prepared by Ptolemy [ca 152; introductory tables], allows a close dating of this record, close enough to make the identification virtually certain. The eclipse in turn, being identified, fixes the chronology of a large part of Assyrian history.

This is an annals record and receives a reliability of 1. It is the oldest astronomical record that is entitled to high confidence, so far as I know. With regard to the magnitude, Fotheringham [1920] argues: "As the eclipse is the only eclipse mentioned in this Chronicle[†], which covers an interval of 155 years, there can be no reasonable doubt that it had been reported as a total eclipse." This is not a safe conclusion. Even in annals the recording of eclipses is highly variable, as Dubs [1938] has shown for Chinese records. Over a span of five centuries, the Anglo-Saxon Chronicles recorded six solar eclipses, of which one (809 Jul 16) was probably far from total. (During this time the Chronicles missed ten or more eclipses that must have been large if not total.) The eclipse of 809, although almost surely not total, was the only eclipse recorded within a span of 145 years. I shall continue to use a standard deviation of 0.06 for the magnitude in the absence of detail indicating a larger magnitude.

[†]Fotheringham means the Assyrian set of annals frequently called the Eponym Canon (which is often written uncapitalized).

The eclipse could have been observed anywhere in Assyria, which will be taken as the rectangle extending from 34°.2 to 37°.4N and from 42°.2 to 45°.2E.

Amos 8:9 refers to an eclipse in terms of prophecy rather than of record. If the verse were inspired by an eclipse, it was probably this one.

-699 Aug 6 BA. Ginzel [1899] discusses at length an Assyrian text that refers to an eclipse. He concludes that the eclipse can probably be identified with the solar eclipse of -699 Aug 6. The place cannot be identified any more closely than somewhere between Assyria and Egypt, and the time is not known more closely than about a century. It cannot be told with assurance if an eclipse of the sun or moon is meant. I have not felt it worthwhile to calculate this eclipse.

Ginzel gives a few other cuneiform inscriptions that refer to lunar eclipses and one that may refer to the solar eclipse of -660 Jun 27. There are a number of brief references from Seleucid times. Many of the eclipses mentioned in these references were not visible in Babylonia or Assyria, so they are probably the results of calculation, perhaps of attempts to predict eclipses. There are extensive calculated ephemerides and "procedure texts" for calculating ephemerides [Neugebauer, 1957]. Ptolemy [ca 152, IV.5] has preserved ten lunar eclipse observations from Babylonia (see Section V.2), and Kugler [1909] has given some more, but apparently no other solar eclipse data are known.

3. ECLIPSE REPORTS FROM CHINA

Table IV.3 is to be used with this section.

61

TABLE IV. 3

LARGE SOLAR ECLIPSES REPORTED FROM CHINA

Identification	Reference	Central Location		Standard Deviations			Reli-ability
		Lat. (deg)	Long. (deg)	Lat. (deg)	Long. (deg)	Magn.	
- 708 Jul 17 C	WG	34.8	114.4	2	2	0	0.1
- 600 Sep 20 C	WG	34.8	114.4	2	2	0	0.1
- 548 Jun 19 C	WG	34.8	114.4	2	2	0	0.1
- 441 Mar 11 C	N	34.8	114.4	3	5	0.01	0.1
- 381 Jul 3 C	NG	34.8	114.4	3	5	0.01	0.1
- 299 Jul 26 C	N	34.8	114.4	3	5	0.01	0.1
- 187 Jul 17 C	D	34.8	114.4	3	5	0	1
- 180 Mar 4 C	DW	34.8	114.4	3	5	0	1
- 146 Nov 10 C	DWG	34.8	114.4	3	5	0.06	1
- 88 Sep 29 C	DG	34.8	114.4	3	5	0.06	1
- 79 Sep 20 C	DW	34.8	114.4	3	5	0	1
- 27 Jun 19 C	DG	34.8	114.4	3	5	0	1
- 1 Feb 5 C	D	34.8	114.4	3	5	0.06	1
+ 2 Nov 23 C	DWG	34.8	114.4	3	5	0	1
65 Dec 16 C	W	34.8	114.4	3	5	0	0.1
120 Jan 18 C	W	34.8	114.4	3	5	0.01	0.1
243 Jun 5 C	W	34.8	114.4	3	5	0	0.1
360 Aug 28 C	WG	34.8	114.4	3	5	0	0.1
429 Dec 12 C	WG	34.8	114.4	3	5	0.01	0.1
1221 May 23a C	W	48.1	114.1	0	a	0.01	1

[a]Cannot be assigned until path direction is calculated. See Section III.1.

References: D = Dubs [1938, 1944, 1955]   N = Needham [1959, p. 420]
G = Gaubil [1732a and b]   W = Wylie [1897]

ca -2000. Reference: Gaubil [1732a, p. 140][†].
"Chung K'ang had just mounted the throne . . . Hsi and Ho, drunk with wine, had made no use of their talents. Without regard to the obligations which they owed the Prince, they abandoned the duties of their office, and they are the first who have troubled the good order of the calender whose care had been entrusted to them: for on the first day of the last

---

[†]Gaubil's French runs: "Tchong-Kang venoit de monter sur le trône . . . Hi, Ho, plongés dans le vin, n'ont fait aucun usage de leur talent. Sans avoir egard a l'obéissance qu'ils doivent au Prince, ils abandonnent les devoirs de leur charge, et ils sont les premiers qui ont trouble le bon order du Calendrier dont le soin leur avoit éte confié: car

moon of autumn, the sun and moon in their conjunction not
being in agreement in <u>Fang</u>, the blind one beat the drum,
the mandarins mounted their horses, and the people ran up
in haste. At that time, Hsi and Ho, like wooden statues,
neither saw nor heard (understood?) anything, and by their
negligence in calculating and in observing the movement of
the stars, they violated the law of death promulgated by
our earlier Princes. According to our inviolable laws,
astronomers who advance or set back the time shall
implacably (or, without pardon) be punished with death. "

The Chinese astronomer's lot was not a happy one.

I find a small problem in the language. Where
Gaubil has "dans leur conjonction n'étant d'accord", which
I have rendered as "in their conjunction not being in agreement", all English translations that I have seen have "did
not meet harmoniously". I presume that all the English
translations trace back to a common source that I have not
found. Neither translation implies an eclipse. The French
says a conjunction, not an eclipse. The English says the
sun and moon did not meet, although perhaps the English
translator meant to say that "they met inharmoniously".
Only an expert in ancient Chinese, with the text before
him, is entitled to guess what the writer meant.

---

au premier jour de la derniere lune d'automne, le soleil
et la lune dans leur conjonction n'étant pas d'accord dans
<u>Fang</u>, l'aveugle a frappé le tambour, les Mandarins sont
montés à cheval, et le peuple a accouru. Dans ce temps-là,
Hi, Ho, semblables à une statuë de bois, n'ont rien vû ni
entendu, et par leur négligence à suputer, et à observer
le mouvement des astres, ils ont violé la loi de mort
portée par nos anciens Princes. Selon nos loix inviolables,
les astronomes qui devancent, ou qui reculent le temps,
doivent être san rémission punis de mort. " The French
text is given in order to let the report discussion rest at
least upon a direct translation and not upon a translation
of a translation.

Almost all students of this passage, including Gaubil, have taken the record to be one of an eclipse. Attempts to identify the eclipse have rested upon the historical period and the reference to Fang. Fang is one of the hsiu, which are collections of stars somewhat analogous to our signs of the zodiac. The Chinese used the hsiu to identify position around the celestial equator while the signs of the zodiac identify position around the ecliptic. Fang is a narrow hsiu covering only a few degrees. Thus solar eclipses visible in China and occurring in Fang, even though probably partial on the basis of the text, are moderately rare. Gaubil [1732a, pp. 141-144] was rather certain that the eclipse occurred on -2154 Oct 12. Other identifications, according to Needham [1959, p. 409], have ranged from -2165 to -1948; Curott [1966] tested -1904 May 12.

If recent conclusions by Needham [1959] and by de Saussure [1930] prove valid, we can perhaps abandon attempts to identify this eclipse. On the basis of their work, I see two likely interpretations of the record; neither makes this an eclipse record.

To start with, it seems that the oldest known text of the passage dates from about the +4th century and that the text has been interpolated into a work otherwise dating from about the -7th century. Needham (p. 189) flatly calls it a forgery. Fotheringham [1958] delicately calls it a "literary restoration, made to take the place of books that were burned by imperial order in 213 B.C., .." Needham (p. 242) says that the famous "burning of the books" is imaginary. Whatever may be the final outcome of the discussion, it is clear that we have little or no knowledge of the history of the record between the event and its writing down 25 centuries later. We can take it as fairly certain that the 4th century "editor" did not understand his material well and that the existing text can be a mixture of history and myth. The passage before us is probably a myth in the same sense that the Arthurian cycle is myth, even though it names a (probably) actual prince.

Needham (p. 188) states that the names Hsi and Ho
of the two astronomers are actually the names of two (or
four or six in some myths) minor solar deities. In the
myths with two Hsi brothers and two Ho brothers, the four
were ordered by the legendary emperor Yao to go to the
four major points of the sun's course (or of the world)
"in order to stop the sun and turn it back to its course at
each solstice, and to keep it going on its way at each
equinox . . . . At the same time, the mythological magi-
cians were charged with the prevention of eclipses." The
activities of the blind one (perhaps a traditional temple
attendant), the horsemen, and the people were probably
prescribed ritual to follow when some mishap overtook the
sun.

We are left with two fragments of fact, the name
Chung K'ang and the hsiu Fang. Chung K'ang [Gaubil,
1732a, p. 142] was the second emperor of the first dynasty,
called the Hsia dynasty (ca -2200 to ca -1800). Since
emperors were commonly identified with gods in ancient
thought [Frazer, 1922], the name Chung K'ang is consistent
with a myth. We are left with Fang.

de Saussure [1930, pp. 102-116] concluded that the
small hsiu† Fang and the adjacent small hsiu Sin originally
made up a single hsiu. This hsiu was originally called Ho.
Around -2000, the hsiu Ho contained the autumnal equinox.
The "last moon of autumn" and the hsiu Ho both place the
terrible misfortune at the time when the sun was at the
autumnal equinox. Further, we learn [Needham, p. 188]
that the station of one of the deities Ho was at the autumnal
equinox. Every fragment of fact has now vanished.

Thus a possible interpretation of the text is that it
is purely mythological, that it prescribes the ritual to be
followed when something goes wrong with the order of the
heavens during autumn, and that it also prescribes the

---

†Which he spelled sieou.

punishment for the deities[†] responsible. Whether the same ritual was to be followed at all seasons or whether there was a separate prescription for each season the text does not tell.

A myth may conceal a fact. If there be a fact behind this myth, I suspect that it concerns time keeping and not an eclipse. The passage emphasizes the damage done to the calendar; Hsi and Ho troubled its good order. The severe but condign punishment was for those who advance or set back the time.

The central problem of ancient calendar makers was that their people needed both the synodic month and the tropical year. The simplest way to predict a calendar is to use cycles, or near-coincidences between some number of months and some other number of years. A cycle that many peoples discovered is the 19-year cycle; this was called the Metonic cycle in ancient Greece and the chang [Needham, p. 406] in ancient China. 19 years = $6939^d.602$ and 235 months = $6939^d.688$. The Gregorian calendar is an elaborate cycle of 400 years; after one cycle, we even come back to the same days of the week.

All cycles will fail when carried far enough. The Gregorian cycle will fail by a day about the year 4500. After 11 of the 19-year cycles, or 209 years, we have 209^y = 76 $335^d.620$ and 2585 months = 76 $336^d.573$. After about this interval, the failure of the cycle becomes obvious. Thus, I suspect that the fact, if any, behind the myth was a failure of the formula being used then to predict the calendar; the gods who pushed the sun along had failed, so that the new moon did not join or pass the sun harmoniously (on the right day?) in Fang or Ho and the calendar was upset[‡].

[†] Many early deities were mortal.

[‡] Pannekoek [1961, p. 87] says that the Chinese year anciently began in the autumn. If this were so at the time of the passage being studied, it implies that a fundamental

I do not interpret the passage as an eclipse report and shall not try to identify it or calculate it.

The Chinese eclipse records to be used in this paper come from court annals, with the exception of the last one. Solar eclipses were believed to be important omens and were noted in the court records. Gaubil [1732b], Wylie [1897], and Hoang [1925] have collected about 1000 solar eclipse records from the annals. I have been able to find only two extensive discussions of these annals. Dubs [1938, 1944, 1955] has analyzed the eclipses reported during the dynasty of the Former (or Eastern) Han, covering the period from about -200 to +23. Gaubil has attempted to cover the entire history of Chinese astronomy up to his own time.

The main problem in using the annals comes in identification. Many of the eclipses listed do not correspond to eclipses in Oppolzer's Canon. In some cases, a simple change corresponding to a credible clerical error brings about correspondence; for these, there is little hesitation in accepting the record. In other cases, several changes must be made in order to yield the date of an actual eclipse. These may represent eclipse reports that have undergone many errors in transcription. If so, identification is uncertain and the records cannot be used with confidence. It is possible that they do not represent actual observations. Eclipses had political importance, and there must have been frequent temptation to invent an eclipse, in spite of the death penalty decreed for such an action. Dubs [1938] believes that a record dated in -183 was invented.

Finally, some of the eclipses may have been calculated. Hoang, in his introduction, points out that some of the eclipse records have appended "This announced eclipse did not occur" to them. This implies that there

epoch in the Chinese calendar was the first day of a year which was also a new moon. This would make the failure of the moon to arrive on schedule serious indeed.

67

was some inaccurate predicting of eclipses, probably by means of cycles. The risk is that a predicted date, which would frequently correspond to an eclipse that occurred somewhere, found its way into the record with the warning left off. The eclipse record just mentioned, which is dated -183 May 6, may be such a case, and the warning may have been omitted for political reasons. There was an eclipse on that date, but it is unlikely that it was visible anywhere in China.

When a record says that an eclipse was total, I have assigned a standard deviation of 0 to the magnitude. When the record does not say total but refers to stars being seen, I have used the conventional value of 0.01. When the record says the eclipse was large, or refers to a hook or crescent shape, I have used the conventional value of 0.06.

In Table IV.3 the column for the local time has been replaced by a column labelled "Reference". This column lists the references used in assigning the standard deviation of the magnitude. It is also the basis for assigning the reliability. I have assigned a reliability of 1 to the records identified and discussed by Dubs and a value of 0.1 to all other records. In doing this, I do not mean to impugn the excellent scholarship of the other sources. The three references to Needham are based upon a passing remark and not upon a discussion by him. Gaubil and Wylie are rather old references. I do not know the extent to which modern investigation would make their work obsolete. Hoang cannot be used for this purpose because he gives no clues to the magnitudes of the eclipses.

Assigning a place to the observations from the annals also causes trouble. The court annals do not give the place of observation. From times long before any date in Table IV.3 there was a court astronomer, but it cannot be assumed that he is the only source of the observations. Some of the records state that the eclipse report came from the provinces, but unfortunately do not state what province. Leaving out the first three reports in

Table IV.3 and the last one, which are special cases, we must take the place to be anywhere in China. Judging from the map given by Dubs [1938], it is reasonable to take China at the period involved as the region from 30° to 40°N and from 106° to 122°E. Kaifeng or Khaifeng (34°.8N, 114°.4E) is near the center of this region and had an observatory, so I have adopted it as the central place for the observations. The standard deviations of 3° in latitude and 5° in longitude correspond roughly to the extent of the region mentioned.

The first three reports are from the annals of the state of Lu (or Lou); Curott [1966] takes the limits of this state to be 34°.5 to 36°.25N and 116° to 119°E. Even though the reports are from state annals, the observations did not necessarily come from the confines of that state. Souciet [1729, pp. 18-20] specifically assigned them to "Caifonfu" (Kaifeng or Khaifeng). I shall follow Souciet in using this as the place of observation, but with standard deviations somewhat smaller than all of China.

There are many Chinese eclipse records after 429. Because of the uncertainty in position, later annals records would have little value, and I have arbitrarily stopped use of the annals at this date.

The eclipse of 360 Aug 28 also appears under the listing 360 Aug 28 M.

The last listing in Table IV.3 needs some discussion.

1221 May 23a C. Reference: Wylie [1897]. In 1221, during the time of the Mongol emperors, a party undertook an expedition across much of central Asia. Wylie translates from the report of this expedition: ". . . at noon, an eclipse of the sun happened, while we were on the southern bank of the [Kerulen] river[†]. It was so dark that the stars could be seen, but soon it brightened up again." At two places that the party reached later,

---

[†]In what we call Outer Mongolia.

69

they compared observations of time and magnitude with those made locally. One of these places cannot be identified. The other place was Samarkand, where they conferred with the local astronomer. The Samarkand observation appears as the record 1221 May 23b C in Section V.3.

This report seems completely reliable. The standard deviation of the magnitude will be taken as 0.01. The only problem concerns the place.

The place can be inferred only from the length of time that the party followed the river and from the day on which they saw the eclipse. Wylie concluded that they were probably at 48°.7N, 116°.25E. Curott concluded that the position was at 47°.5N, 112°E. These seem like the extreme possibilities. I shall take these positions as defining the limits of the line on which the observation was made, and shall assign 48°.1N, 114°.1E as the central place.

4. ECLIPSE REPORTS FROM EUROPE

Table IV.4 is to be used with this section.

Ginzel [1899] lists several references to solar eclipses in the writings of Livy. The dating is usually vague. Sometimes the dating is detailed enough to make us sure that there was no corresponding large eclipse. More than half of the eclipses were accompanied by a rain of stones (hailstorm?). I presume that all these reports are of magical eclipses.

Ginzel also lists what he calls "doubtful eclipses" described by the writer Julius Obsequens. I pass by these in silence.

-50 Mar 7 E (Caesar). Reference: <u>Dio Cassius</u> [ca 230, Bk. xli, Chap. 14]. There are both modern and ancient mysteries connected with this eclipse.

TABLE IV. 4

LARGE SOLAR ECLIPSES REPORTED FROM EUROPE

Identification		Hour if Given	Central Locations		Standard Deviations			Reliability
			Lat. (deg)	Long. (deg)	Lat. (deg)	Long. (deg)	Magn.	
- 50 Mar 7	E		43.0	13.0	0	a	0.06	0
59 Apr 30	E	13 1/2	40.8	14.5	0	a	0.1	1
402 Nov 11	E		40.0	-5.0	1.7	2.3	0.06	1
418 Jul 19	E		40.0	-5.0	1.7	2.3	0.06	1
447 Dec 23	E		40.0	-5.0	1.7	2.3	0.06	1
538 Feb 15	E	08	44.4	12.2	0	10.0	0.06	1
540 Jun 20	E	09	44.4	12.2	0	10.0	0.01	1
634 Jun 1	E		46.0	2.0	1.7	2.3	0.06	0.005
779 Aug 16	E		46.0	2.0	1.7	2.3	0.06	0.005
1030 Aug 31	E		59.0	13.0	2.9	4.0	0.06	0.5
1185 May 1	E		55.0	10.0	2.9	2.3	0.06	0.5
1239 Jun 3a	E	13 1/2	43.6	3.9	0	0	0.01	1
1239 Jun 3b	E		43.7	5.6	0	0	0.01	1
1239 Jun 3c	E		44.1	6.2	0	0	0.01	1
1239 Jun 3d	E		44.4	8.9	0	0	0.01	1
1239 Jun 3e	E	13 1/2	44.9	8.6	0	0	0.01	1
1239 Jun 3f	E		45.5	9.2	0	0	0.01	1
1239 Jun 3g	E	15	45.0	9.7	0	0	0.01	1
1239 Jun 3h	E	15	44.8	10.3	0	0	0.01	1
1239 Jun 3i	E		43.8	10.5	0	0	0.01	1
1239 Jun 3j	E	15	44.7	10.6	0	0	0.01	1
1239 Jun 3k	E	13 1/2	44.6	10.9	0	0	0.01	1
1239 Jun 3ℓ	E	15	43.8	11.3	0	0	0.01	1
1239 Jun 3m	E	15	43.3	11.3	0	0	0	1
1239 Jun 3n	E	12	43.5	11.9	0	0	0	1
1239 Jun 3o	E	15	44.4	12.2	0	0	0.01	1
1239 Jun 3p	E	15	44.1	12.2	0	0	0.01	1
1241 Oct 6a	E	>12	51.1	13.3	0	0	0.01	1
1241 Oct 6b	E	15	49.3	12.0	0	0	0.01	1
1241 Oct 6c	E		49.0	10.2	0	0	0.01	1
1241 Oct 6d	E		48.8	10.3	0	0	0.01	1
1241 Oct 6e	E		48.8	13.1	0	0	0.01	1
1241 Oct 6f	E		48.4	10.9	0	0	0.01	1
1241 Oct 6g	E		48.3	13.4	0	0	0.01	1
1241 Oct 6h	E	15	48.3	16.3	0	0	0.01	1
1241 Oct 6i	E		48.3	11.7	0	0	0.01	1
1241 Oct 6j	E	15	48.2	16.4	0	0	0.01	1
1241 Oct 6k	E	15	48.1	13.9	0	0	0.01	1
1241 Oct 6ℓ	E		47.8	13.0	0	0	0.01	1
1241 Oct 6m	E	15	47.6	14.5	0	0	0.01	1
1241 Oct 6n	E		48.4	11.7	0	0	0.01	1

[a] Cannot be assigned until path direction is calculated. See Section III. 1.

The following passage [Chambers, 1915] illustrates the modern mystery: "An eclipse of the Sun is mentioned by Dion Cassius as having happened when Caesar crossed the Rubicon. ...There seems no doubt that the passage of the Rubicon took place in 51 B. C., and that the eclipse must have been that of Mar 7, 51 B. C."[†] Chambers then cites Dio (or Dion) Cassius, Book xli, Chapter 14.

The mystery starts with the discovery that the cited chapter does not deal with the crossing of the Rubicon. On the assumption that the chapter number is a misprint, we go back to Book xli, Chapter 4, which does deal with the Rubicon crossing. There we find: "When Caesar was informed of this[‡], he came to Ariminum[*], then for the first time overstepping the confines of his own province, ..." (Translation mentioned in the reference.) No mention of an eclipse of the sun.

Plutarch [ca 100, Life of Caesar] gives a little more detail. According to Plutarch, Caesar crossed the Rubicon at night, which makes it hard to see a solar eclipse.

Book xli, Chapter 14 of Dio does mention an eclipse of the sun. It mentions it in connection with the flight of Pompey from Brundisium (Brindisi) across the southern Adriatic to Dyrrachium (Durres or Durazzo). It is one of about a dozen portents that happened during his crossing or his landing or during that year. The passage does not seem worth quoting here, although it might interest a student of magic. There is a slight implication that the

---

[†] There is almost every doubt that the crossing occurred in 51 B. C. (-50). It happened in the early part of -48, on the basis of historical evidence that I have not seen questioned seriously.

[‡] A Senate decree ordering him to disband his army.

[*] The modern Rimini.

eclipse was placed at Rome. The eclipse is magical and has no reliability.

Pompey's crossing is dated rather accurately, and no significant eclipse was visible in Italy during that year. The eclipse of -50 Mar 7, more than two years before, almost surely crossed part of Italy and may have been assimilated to the event. If so, we have no clue as to where it was seen. In case the reader should wish to use this eclipse report, I shall calculate it, with a reliability of zero, a standard deviation of magnitude of 0.06, and a place anywhere in Italy. It is convenient to replace Italy by the line running from 46° N, 9°E to 40°N, 17°E in the calculations.

Among the recent discussions that I have seen of this eclipse, only Ginzel [1899] correctly quotes Dio. Ginzel calls this one of the "alleged eclipses from the time of Caesar". He quotes several passages connected with Caesar that describe some sort of darkness. One is the passage by Plutarch quoted in Section III.2; another is the passage by Virgil mentioned in connection with the eclipse report 1133 Aug 2b B. A quotation from the poet Lucan (Pharsal. I, 540) comes closest to placing an eclipse at the Rubicon crossing. Lucan has it occur during Caesar's advance on Rome after the crossing.

59 Apr 30 E. Reference: Ginzel [1899]. Ginzel gives references to this eclipse from several ancient writers. Pliny the Elder (Natural History, Bk. II, 180) gives the most detail. The eclipse was observed in Campania between the 7th and 8th hours of the day, and in Armenia between the 10th and 11th hours. The Armenia observation is treated as the separate observation 59 Apr 30 M.

Pliny was a serious student of natural history among other things, and was killed on 79 Aug 25 while trying first to observe the great eruption of Vesuvius and later to rescue some friends [Pliny (the Younger), ca 100, Letter LXV to Tacitus]. He is often regarded as somewhat gullible and willing to accept the marvelous. This does not seem to warrant questioning the eclipse report, which he probably wrote close in time to the event, and in which he seems to have been interested in the difference in times between Armenia and Campania. I give this record a reliability of unity. The standard deviation of magnitude will receive the value 0.1 because eclipse predictions were rather reliable by now and may have been known to a number of people.

Campania was a rather narrow region running from northwest to southeast. It is conveniently replaced by the line running from 41°N, 14°E to 40°.5N, 15°E.

The identification of the eclipse seems secure.

The other Italian reports of the eclipse can be ignored. Dio Cassius [ca 230, Bk. lxi, Chap. 16] shows assimilation at work by claiming, two centuries later, that the eclipse was absolutely total and that stars were seen at Rome during the funeral services of Agrippina, the notorious mother of Nero and the alleged murderess of her uncle and husband the Emperor Claudius.

186 Dec 28 E. Reference: Ginzel [1899]. Ginzel quotes a passage from the 4th century historian Lampridius who says that an eclipse was seen in Rome about Jan 1 during the reign of Commodus (180-192). There was only one eclipse that can meet these conditions. Unfortunately, the sun must have set at Rome while still eclipsed. Under these circumstances, even a small eclipse would be easily seen, and there is no clue to the magnitude. This eclipse will not be calculated.

Ginzel [1899] discusses a large number of eclipse reports from chronicles or historians of the late Empire. Most of the reports are dated fairly well and the corresponding eclipses can be identified. Unfortunately the place can usually be assigned no more closely than to most of the Roman Empire. Such reports have negligible value for the purposes of this paper, although they may help the historian to confirm or to refine the chronology of the late Empire. There are a few exceptions in which the place is known well enough to give the report a marginal value. Ginzel will be the reference for the reports from Europe until the report 538 Feb 15 E.

402 Nov 11 E. This report is from the annals of Bishop Hydatius, who had his bishopric in northern Portugal but who used Spanish as well as local records: "0ℓ. 295, 2. The sun was eclipsed on the third of the Ides of November on the second week-day." The date translates into 402 Nov 11, which was a Tuesday[†]. Reliability = 1; standard deviation in magnitude = 0.06. The place is the Iberian peninsula, which can be taken as the region from 37° to 43°N and from 1°W to 9°W. It is interesting that Hydatius still used the Olympiads and not the Christian era in his dates.

418 Jul 19 E. Hydatius also reports this eclipse, which was reported independently from the other end of the Empire (418 Jul 19 M): "0ℓ. 299, 2. The sun was eclipsed on the 14 Kalends of August, which was the fifth week-day." The date translates into 418 Jul 19, which was a Friday. This report has the same characteristics as the report 402 Nov 11 E.

447 Dec 23 E. By Hydatius: "0ℓ. 306, 3. Yr. 23, Valentinian III. The sun was eclipsed on the 10 Kalends[‡] of January, which was the third week-day." If "10 Kalends" is right, the date equals 447 Dec 23, a Tuesday. This report receives the same characteristics as the last two. Valentinian III was Emperor of the West from 425 to 455.

[†] In the records 402 Nov 11 E and 418 Jul 19 E, the week-day or feria is given as if Monday, not Sunday, were the first feria. I have found no other indication of such a convention. In his other records that I noted, Hydatius used the ordinary convention.

[‡] Some copies have "9 Kalends".

For 458 May 28, Hydatius lists a partial solar eclipse with an estimate of the magnitude; the sun was diminished until it appeared like a moon on its fifth or sixth day. While a partial eclipse with a measured magnitude is as good as a total eclipse, this measurement does not seem accurate enough to be valuable, especially since the place of the observation is uncertain. Hydatius also lists a lunar eclipse on 462 Mar 2.

538 Feb 15 E. This record is best discussed along with the record for 540 Jun 20 E.

540 Jun 20 E. References: Ginzel [1899]; Bede [731, Bk. V, 24]. Ginzel quotes a passage from a document found in Ravenna that I have not seen directly: "Fourth year of Belisarius and fourth year of Straticus, there was darkness from the third till the fourth hour of the day on Saturday." Bede writes: "In the year 538, there happened an eclipse of the sun, on the 16th of February, from the first to the third hour." "In the year 540, an eclipse of the sun happened on the 20th of June, and the stars appeared during almost half an hour about the third hour of the day." The quotations are from the translation mentioned in the citation.

These are two of three solar eclipses recorded by Bede. The third appears under the listing 664 May 1 B.

The statements are taken from annals and may be given a reliability of unity. Neither document states anything specific about the magnitude of the eclipse of 538, so I give it a standard deviation of 0.06 in magnitude. "Almost half an hour" sounds a little long for stars to remain visible, except possibly for Venus, but time-keeping was rather informal then and this duration can be ignored. On the basis of the stars, I give the eclipse of 540 a standard deviation of 0.01 in magnitude. The main problem concerns the place of these observations.

Bede was writing an ecclesiastical history of England and drew mostly upon local sources for the history before his own time. However, he includes some general history and for this he used Continental sources. It is quite possible that he did not realize how narrow the zone of a total solar eclipse is and hence did not realize that a Continental eclipse observation does not necessarily apply to England.

The charts in both Oppolzer's and Ginzel's canons of eclipses show that the eclipse of 538 Feb 15 passed through the eastern Mediterranean and that the eclipse of 540 Jun 20 passed along the Mediterranean and the southern part of Europe. It is unlikely that either eclipse, and particularly the one in 538, would have been reported from England, and it is most unlikely that stars would have been visible in England during the 540 eclipse. Further, the conditions of life in England at the time were unfavorable to making and recording such observations; the first official church presence in England dates from 597. In spite of this, I would have assigned the places of observation to be somewhere in England if I had not found the Ravenna document.

The Ravenna document gives the only clue we have about a place of observation, and it is not satisfactory. There is one report but there were two eclipses. According to Ginzel, the listing of Belisarius and Straticus as consuls identifies the year as 539. This does not agree with either eclipse. "Saturday" does not help because it too does not agree with either eclipse. Since there is no way to choose, Ginzel labels the Ravenna record as "538 or 540". It seems possible that the compiler of the Ravenna record accidentally ran two records together and that the single record does in fact refer to both eclipses, but there is no evidence that this actually happened.

The Ravenna record does nothing more than establish the fact of a Continental record for at least one of the eclipses of 538 and 540, and hence it gives credibility to a non-English source for Bede's reports. The Ravenna record is certainly not the record used by Bede[†]. If it refers to both eclipses, it is almost surely not an original record. However, it is most unlikely that the eclipse reports originated from England, and Ravenna (44°.4N, 12°.2E) is the only place we have, so I shall use it. I shall indicate lack of faith in the choice and keep it from having much importance by assigning a standard deviation of 10° to the longitude.

Attribution of a Continental source for these records does not imply that Bede traveled there; that is a separate question.

634 Jun 1 E. See 779 Aug 16 E.

779 Aug 16 E. Reference: Turoldus [ca 1100]. Lines 1423-1437 of the reference can be translated approximately as follows: "In France there was great and marvelous trouble: a storm of wind and thunder, of immeasurable rain and hail; lightning strokes came close and fast, and there was a truly great earthquake. From Mont St. Michel to Seinz[‡], from Besançon to the port of Wissant, there was not a house whose walls did not burst. About noon there was a great darkness; there was light only when the heavens were rent. None saw it without terror. Many said 'It is the end of the world come upon us'. They did not know nor speak truly: It was the great mourning for the death of Roland."

As an eclipse report, this description of tempest, hail storm, earthquake, and darkness resembles passages from Livy, or the report by Dio Cassius of the alleged eclipse connected with Pompey (see -50 Mar 7 E). Only if we can establish probable assimilation of an actual eclipse can we give this passage astronomical value in addition to its poetic value.

[†]Since Bede had records that clearly identified both eclipses, and with detail that is not in the Ravenna record.

[‡]This is the spelling of the original. Seinz is probably the modern Saintes in the Cognac region. Saintes and the other points named are at the corners of a large parallelogram.

In the summer of 778 [Holland, 1910, for example],
Charlemagne campaigned in Spain. On 778 Aug 15, as he
was returning to France, he was ambushed somewhere in
the western Pyrenees and his rear guard was almost destroyed. He lost several important officers in this action;
among them was "Hruodland, praefect of the Breton march".
This reference to Hruodland is apparently the entire basis
for an historical Roland. The date given for the action
seems secure. The eclipse nearest in time to this date
that would have been large in France or northern Spain was
the eclipse of 779 Aug 16, and it is possible that this eclipse
was assimilated to the action. An error in dating that would
allow the action to occur on the date of the eclipse but to be
recorded as 778 Aug 15, whether by confusion on the part
of the historian or by copying error, is unlikely since it
would mean two independent errors in writing the date of
the battle.

This is not the only possible identification. As an
unsigned article [Ency. Brit., 1911] points out, the historical rear guard action may not have been the battle involved
in the original form of the legend. The action of 778 Aug
15 affords no basis for the legend of the twelve paladins or
peers who had a prominent part in the legendary battle.
Further, the scale of the rear guard action is not comparable with the importance given the battle in the legend, although normal exaggeration could account for this discrepancy. However, in the years 636-7, there was a Frankish
campaign in Spain led by 12 Frankish chieftains. Several
of them were killed in a battle in the western Pyrenees.
This battle, which may be legendary, could be the one in-
corporated into the earliest form of the Roland legends. If
an eclipse were assimilated to this battle, it would almost
surely be the eclipse of 634 Jun 1.

The reliability of the assumption that the passage
from La Chanson de Roland is related to an actual eclipse
is small but finite, say 0.01. This value will be divided

equally[†] between the two possibilities. It should be noted that the passage, like many other passages considered, refers only to a great darkness and does not mention an eclipse specifically. For this reason, and because of the large time difference between either battle and either possible eclipse, the reader may wish to use a smaller reliability. However, the reliability of this report seems to me to be comparable with that of the eclipse of Archilochus (see -647 Apr 6 M) or the eclipse of Thales (see -584 May 28 M), to name but two examples.

The passage makes the darkness great, so that the magnitude of the eclipse, if an eclipse is meant, would be taken as unity if the passage is taken literally. However, there is certainly exaggeration in the passage, and no distinctive features of a total eclipse are mentioned, so that the standard deviation of the magnitude will be taken as 0.06.

The place of the observation will be taken as anywhere in France, although plausible arguments can be given [Ency. Brit., 1911] that the Roland legends arose in or near Brittany. France can be taken as equivalent to the region from 43° to 49°N and from 2°W to 6°E. It is interesting that the eclipse and the other signs were not at the same place as the battle.

Because of low reliability, and because of the poor identification in both time and place, this eclipse report contributes little in the way of useful data; its main value for the purpose of this paper is indirect. It gives a "textbook" example of the magical eclipse and furnishes a useful standard in the assignment of low reliability to eclipse reports that have heretofore been taken at face value. It is unique in that it states a magical reason for the darkness

_____

[†] There is a slight argument in favor of 779 Aug 16. This eclipse came almost exactly on the first anniversary of the battle of 778 Aug 15. If the weather allowed it to be observed, it must have made a strong impression on those with memories of the battle.

and other calamities. The reason is not the one given in Section III. 2, but, for a writer who had lost knowledge of the original significance of a magical eclipse, the step from the original reason to the one given is small and natural.

1030 Aug 31 E (Stiklestad). Reference: Sigvat Skald quoted by Snorri [ca 1230]. In the summer of 1030, Olaf Haraldsson, the once king and future saint of Norway, was killed in a battle at the place called Stiklestad while trying to regain his throne. The contemporary scald Sigvat commemorated the event in a poem that survives only in fragmentary quotations. In the translation cited for Snorri in the references, the relevant passage (from "St. Olaf's Saga", Chap. 227) reads:

> No small wonder, say the
> sailship-steerers, was it,
> when from cloudless heaven
> hardly warmth gave the sun-orb.
>
> An awful omen - from the
> English I learned the portent -
> for the king that fast did
> fail daylight in battle!

Apparently this passage furnishes the only support for the claim that the eclipse of 1030 Aug 31 happened during and at the battle of Stiklestad, and this support is not strong. All other evidence puts the battle on 1030 Jul 29. Before turning to the other evidence, the problems concerning Sigvat's passage will be discussed.

The eclipse has usually been called the "eclipse of Stiklestad", but it would be more meaningful to call it the "eclipse of Olaf" since it allegedly occurred at his death. It is one of many miracles or marvels associated with him. Sigvat [Snorri, St. Olaf's Saga, Chap. 245] describes two others. A passage in the same chapter quoted from another contemporary poet, Thorarin Loftunga, describes still others and shows that the cult of Olaf was already well established when he wrote. The eclipse described by Sigvat is therefore clearly a magical eclipse in the context where it is found, like the "eclipse of Roland" (see report 779 Aug 16 E) or some of the "eclipses" associated with Caesar (see report -50 Mar 7 E).

That is, the eclipse is magical if the quoted reading of the passage is adopted. The passage is singularly difficult[†]. Many interpretations of it have been suggested. The translator cited clearly accepted the interpretation that puts the eclipse at the battle. Snorri [ca 1230] also clearly accepted this interpretation. Snorri was as skillful as Herodotus (see -477 Feb 17a M) in imagining how an eclipse might have affected events. Landmark [1931] discussed other interpretations; the problems in deciding what the passage means go well beyond mere translation.

The first half of the passage is essentially the same, so far as our purposes are concerned, in most interpretations. Landmark gives a different translation[‡] of the second half: "Heavy was it on the day of the king's sign - daylight reached not its beautiful lightness - I learned from the east about the battle's course." The reader should notice that this translation separates the sign from the battle, and that it does not imply that an eclipse occurred at either the

---

[†]In the edition cited the original language is, using strokes to separate lines: Undr láta þat ýtar / eigi smátt, es máttit / skaenjorðungum skorðu / skýlauss roðull hlýja; / drjúg varð á því doegri / (dagr náðit lit fogrum), / orrostu frák austan / atburð, konungs furða. "Austan" usually means "from the east" although it is used on rare occasions [Kratz, 1969] to mean "in the east". This is the term that appears as "from the English" in the translation quoted above. It seems to be the scholarly consensus that "east" rather than "English" is meant.

[‡]Landmark translated Sigvat into modern Norwegian. I am indebted to Mr. Sverre Kongelbeck of the Applied Physics Laboratory for translating the needed parts of Landmark's paper and of Snorri into English for me. Kratz [1969] prefers to read the first line as "Great (or remarkable) was the king's sign that day."

time or place of Olaf's death. The eclipse was merely a sign, like the sign associated with Magnus Erlingsson (1185 May 1 E) or with Proclus (484 Jan 14 M).

Two interpretations associate the eclipse with Sigvat's hearing about Olaf's death, not with the death itself. Sigvat [Snorri, ca 1230, Saga of Magnus the Good, Chaps. 7 and 8] was not in Norway when Olaf was killed. He felt that he had been reproached for his absence and that he needed to explain it. He said that he had been on a pilgrimage to Rome and that he heard about the battle and Olaf's death on his way home. He said, apparently as a description of his reaction to the news (in the cited translation): "On the Mont I stood, remembering / many targes sundered. .", and so on.

Since the direct route from Rome to Norway passes through the Alps, one interpretation holds that Sigvat heard the news while he was in the Alps, and that "Mont" refers to some point in them. Another interpretation points out that the most specific statement about the darkness is that the "sailship-steerers" experienced it. On this basis it is suggested that the news-bearers saw the eclipse on the sea-borne part of their trip.

If leading scholars can produce interpretations as different as these, I am not competent to judge among them. However, I must either ignore this record or make a choice among competing interpretations. It seems to me that the interpretation involving the Alps requires more assumptions than the others to make it work, so I shall reject it. I shall assume that the eclipse was seen by the "sailship-steerers". This term could be a reference to people on a ship; in this case, the eclipse was probably seen in the Baltic. However, [Kratz, 1969], the sea was intimately associated with Norse life and warfare and the term may simply mean "the people" or "the warriors" and may not specifically mean people on a ship. In this case, the eclipse was probably seen in Norway. [†]

[†] This does not specifically account for "from the east", but there is no need to. Perhaps it means that the news reached

Whatever interpretation is accepted, Sigvat's passage gives little support to the claim that the eclipse of 1030 Aug 31 happened at Olaf's death. Either it was not associated with his death or it was magical if it were. Experience with other alleged eclipses shows that one cannot safely accept a magical eclipse without independent supporting evidence. Here, all independent evidence denies the possibility that the eclipse happened at Olaf's death.

Snorri [ca 1230, St. Olaf's Saga, Chap. 235] claimed that the eclipse (or darkness; the language apparently did not make this distinction at the time) happened during the battle, and then, perhaps unwittingly, he contradicted himself by saying that the date was Jul 29. Earlier evidence for the date comes from the fact that Jul 29 is St. Olaf's Day; this date was established when Olaf was canonized in 1164, and it was chosen to commemorate the date of his death. However, the cult of St. Olaf existed long before his canonization, and the earliest evidence about the date of his death comes from early statements giving the date celebrated, with accompanying statements that the date commemorated the date of his death. Early written evidence of the date has been discussed by Landmark [1931] and by Dickins [1940]; it goes back to about 1050. There is evidence that celebration of his cult goes back even farther, and hence that the existing records derive the date from still older sources. If we can accept an account by Snorri [Saga of Magnus the Good, Chap. 10], Olaf's son Magnus built a shrine to Olaf, apparently about the time of his accession in 1037. Also, according to Snorri [St. Olaf's Saga, Chap. 244], the first service in honor of Olaf was on 1031 Aug 3, which was stated to be a year and five days after his death. On this occasion his body was moved to a church in Nidaros (the present Trondheim), and another miracle occurred. We also learn from Snorri [Saga of the Sons of Magnus, Chap. 30] that Aug 3 was called the "latter day of St. Olaf's Mass."

Sigvat by way of, for example, Russia. Perhaps an Icelandic usage has prevailed and "from the east" means "from Norway".

In summary, there are written records going back to about 1050 which give Jul 29 as the date of Olaf's death. There is strong evidence that these records continue a still older tradition. No old document gives any other date. However, there is no strictly contemporary record of the date. Olaf died on Jul 29 beyond a reasonable doubt but not beyond all unlikely possibilities.

I have not seen any indication that belief in the "eclipse of Stiklestad" was part of the early body of belief about St. Olaf. The first known connection of the eclipse with Olaf's death seems to be Snorri's interpretation of Sigvat's poem. If this be so, the eclipse is not assimilated[†].

In fact, on the basis of the evidence, the eclipse is neither the "eclipse of Stiklestad" nor the "eclipse of Olaf". It is the eclipse of Sigvat. [‡] Viewed in this light, it may be literary; Sigvat may have thought that an eclipse was an appropriate sign. Since there was a known eclipse very near the time and place in question, I shall assume that Sigvat's eclipse is the eclipse of 1030 Aug 31. With a reliability that I shall take to be 0.5, I shall assume that he

---

[†]Neglecting the fairly small possibility that Sigvat himself did the assimilating. Landmark [1931] pointed out that the eclipse of 1124 Aug 11 came close in time to St. Olaf's Day and that it passed through Norway. He suggests that it may be the eclipse, if any, that was assimilated to Olaf's death in popular tradition. The tradition of this eclipse may in turn have lead to Snorri's interpretation of Sigvat. However, we do not know enough about the early history of Olaf's cult to know when the eclipse tradition became part of it, and the eclipse of 1030 Aug 31 may well be the one assimilated to Olaf's death if there were one.

[‡] To confuse matters further, there is no certain evidence [Kratz, 1969] that Sigvat or any other one person composed all the passages that Snorri attributed to him. For simplicity, I shall continue to write as if Sigvat were the composer.

meant to place the sighting of the eclipse[†] in Norway or at sea. The sea could have been the Baltic[†] between the Scandinavian peninsula and the German or Danish or Russian coasts. The region between latitudes 54° and 64°N and between longitudes 6° and 20°E seems a reasonable one to use.

"Lack of warmth" gives the only clue to the magnitude. This may have been a literary detail furnished by Sigvat. In the absence of definite information, I shall use the conventional value of 0.06 for the standard deviation of the magnitude.

1185 May 1 E. Reference: Storm [1888]. Storm has edited ten Icelandic annals. The annals show considerable interdependence, and it is hard to decide how many independent sources recorded a particular item of history. Many of the annals have an entry for the year 1184 that can be translated as: "Fall of King Magnus Erlingsson; a sign in Oslo; darkness over the southern lands." One of the annals does not contain the reference to Oslo. One (the Annales regii) inserts a reference to the death of "Heinrekr ungi Englakonungr" between the reference to Magnus and the "sign of Oslo"; this Henry is probably Prince Henry, heir apparent to Henry II of England, who died in 1183. Although the matter is not certain, I shall take it that the annals tell us of a sign in Oslo, that this sign was an eclipse which also caused "darkness over the southern lands", and that the sign[‡] was associated with the fall of King Magnus.

[†]Originally I accepted "from the English" as a possible reading and included the North Sea as well as the Baltic. Because of the low statistical weight attached to this eclipse, the change has trivial consequences.

[‡]It is interesting that the sign was in Oslo, while Magnus was killed on the west coast. The sign and the death have not been made to coincide in place; it is not clear whether they are meant to coincide exactly in time or not.

The identification and the reliability pose a problem similar to the one we had with the record 1133 Aug 2a B. King Magnus V Erlingsson was defeated and killed on 1184 Jun 15, about a year away from the only possible eclipse.[†] Simple assimilation is probably the explanation, but because of the dating discrepancy and the slight magical implication, I shall lower the reliability to 0.5. The standard deviation of the magnitude will receive the conventional value of 0.06.

In the context of this report, the "southern lands" may be the southern parts of Norway. With perhaps greater probability, they may be the parts of Europe[‡] lying across the Baltic. The probability that the annalist would have heard of the eclipse decreases with increasing distance from Oslo. Hence I shall take the region of observation to stretch from Oslo southward to the Frankfurt-Prague line; the choice of this line is clearly arbitrary. Use of the regions between latitudes $50°$ and $60°N$ and between longitudes $6°$ and $14°E$ seems reasonable.

1239 Jun 3 E. Reference: <u>Celoria</u> [1877a]. Celoria gives about 35 reports concerning this eclipse. All are simple statements from the annals of various towns or monasteries. It seems unnecessary to quote them. After we eliminate those that seem to be copies and those that give no indication of near-totality, 16 remain. These are listed in Table IV.4 under the identifications 1239 Jun 3a E

[†]The fact that the only possible eclipse came after the events noted poses no problem. It is clear that the annals were written, at least in part, after the events and not strictly contemporaneously with them. This is shown, for example, by the custom of recording the length of a pope's reign in the year of his election, not of his death.

[‡]I thank Mr. Kaye Weedon of Blommenholm, Norway, for this suggestion. I did not include this part of Europe in the original version of Part I mentioned in the Preface. The change does not affect the values found for the accelerations; this eclipse cannot be used, for a technical reason explained in Section XIII.2.

through 1239 Jun 3p E. The places that correspond to the letters are as follows: a, Montpelier; b, Mirabeau; and c, Digne in France; d, Genoa; e, Alessandria; f, Milan; g, Piacenza; h, Parma; i, Lucca; j, Reggio; k, Modena; l, Firenze; m, Siena; n, Arezzo; o, Ravenna; and p, Cesena in Italy.

These reports pose a problem that has not arisen with reports discussed earlier. Earlier reports were observed by a particular person or at a particular place. Selection of the eclipses reported was governed by the accidents of which eclipses were observed or which reports were preserved. There is no reason to think that the process of selection was biased in any way that is significant for the purposes of this study. In contrast the reports of the eclipse of 1239 Jun 3 are all those that have been found for that eclipse, and we must ask whether there is a likely bias in their distribution. The answer is yes.

The reports listed came from either France or Italy. The path of the eclipse was almost parallel to the Mediterranean coast of France and was probably somewhat to the south of the coast. All reports near this portion of the path almost surely came from points north of it; there could not have been any reports from the south unless the curve in Oppolzer's Canon is seriously in error.

Reports from Italy could have come from points either north or south of the path. However, it is my impression that the region to the north was much more populated than the region to the south. If so, reporting from the north side was more likely in Italy just as it was in France.

Thus there is a strong suspicion that the records are biased, for geographical reasons, toward the north of the eclipse path. Accordingly I shall not use the reports of the eclipse of 1239 Jun 3 although I shall carry out the calculations for them[†].

1241 Oct 6 E. Reference: <u>Celoria</u> [1877b]. Celoria gives similar reports concerning this eclipse. Those that explicitly indicate a large magnitude and that do not seem to be copies are listed in Table IV. 4 under the identifications 1241 Oct 6a E through 1241 Oct 6n E. The places that correspond to the letters are as follows: a, Altenzelle; b, Ensdorf; c, Ellwangen; d, Neresheim; e, Altaich; f, Augsburg; and g, Richersberg in Germany; h, Neunburg, Austria; i, Scheftlarn, Germany; j, Vienna; k, Lambach; ℓ, Salzburg; and m, Admont in Austria; and n, Weihenstephan, Germany.

We must ask whether there is a likely bias in the distribution of these places. According to the chart in Oppolzer's <u>Canon</u>, the eclipse path cut across Europe from northwest to southeast, approximately from Hamburg to Istanbul. The question concerns a possible bias in the east-west distribution of the reports.

The places listed lay, at the time of Celoria's work, in either the German Empire or the Austro-Hungarian Empire. There are many places in both empires on both sides of the path. Since Celoria had access to the records cited, he should also have had access to the records of other places in both empires. Thus it is reasonable to assume that there is no bias from the standpoint of Celoria's access to the records.

The political situation in 1241 was different from what it was in 1877. The path lay close to the eastern bound of the Holy Roman Empire, and all the places listed lay within that Empire. We must consider whether this distribution was caused by the position of the path or by social and political conditions.

[†]In preparing the original version of Part I, I had not realized the possibility of a bias and I originally intended to use these reports. As matters turned out, I could not have used them anyway, for a technical reason that will be explained in Section XIII. 2.

The kingdoms of Poland and Hungary lay east of the Holy Roman Empire. If records of the eclipse were made at points much to the east of any listed, they would have been made in one of those kingdoms. Both kingdoms were as likely to be sources of eclipse reports as the Empire except for one factor.

There was a large Mongol invasion of Europe in 1241. Poland and Hungary were hit heavily and at the time of the eclipse much of Hungary was under a Mongol occupation that lasted for about a year. It is commonly said [Macartney, 1958, for example] that the invasion and occupation caused frightful devastation in Hungary. Since the occupied people would have a natural inclination to exaggerate conditions, it is hard to judge how much devastation and depopulation actually occurred. Certainly it is reasonable to assume a decrease in the number of places where records would be made and preserved in a stable environment.

However, there is a compensating factor. The people who were left to observe would be, it seems to me, more likely than usual to take an eclipse as a sign or token of their misfortune and consequently more likely to record it. Thus, there may have been fewer people to observe but those few may have been more likely to record.

In the absence of detailed information I shall assume that the factors affecting the likelihood of record-making balanced each other. Consequently, I shall assume that the reports of the eclipse of 1241 Oct 6 have no geographical bias.

5. ECLIPSE REPORTS FROM MEDITERRANEAN COUNTRIES

Table IV.5 is to be used with this section.

-688 Jan 11 M. See -647 Apr 6 M.
-661 Jan 12 M. See -647 Apr 6 M.
-660 Jun 27 M. See -647 Apr 6 M.
-656 Apr 15 M. See -647 Apr 6 M.
-647 Apr 6 M (Archilochus). References: Ginzel
[1899]; Fotheringham [1920]. "Nothing there is beyond
hope, nothing that can be sworn impossible, nothing won-
derful, since Zeus father of the Olympians made night
from midday, hiding the light of the shining sun, and sore
fear came upon men." The translation is by Fotheringham.

TABLE IV. 5

LARGE SOLAR ECLIPSES REPORTED FROM MEDITERRANEAN COUNTRIES

Identification	Hour if Given	Central Location		Standard Deviations			Reliability
		Lat. (deg)	Long. (deg)	Lat. (deg)	Long. (deg)	Magn.	
- 688 Jan 11 M		39. 0	24. 9	1. 1	0	0. 06	0. 01
- 661 Jan 12 M		39. 0	24. 9	1. 1	0	0. 06	0. 01
- 660 Jun 27 M		39. 0	24. 9	1. 1	0	0. 06	0. 01
- 656 Apr 15 M		39. 0	24. 9	1. 1	0	0. 06	0. 01
- 647 Apr 6 M		39. 0	24. 9	1. 1	0	0. 06	0. 01
- 634 Feb 12 M		39. 0	24. 9	1. 1	0	0. 06	0. 01
- 607 Feb 13 M		39. 0	31. 0	1. 2	2. 3	0. 3	0. 01
- 587 Jul 29 M		39. 0	31. 0	1. 2	2. 3	0. 3	0. 01
- 584 May 28 **M**		39. 0	31. 0	1. 2	2. 3	0. 3	0. 01
- 581 Sep 21 M		39. 0	31. 0	1. 2	2. 3	0. 3	0. 01
- 487 Sep 1 M		38. 3	23. 3	2. 0	2. 0	0. 06	0
- 477 Feb 17a M	Morn. ?	38. 5	28. 0	1. 0	1. 0	0. 06	0. 1
- 477 Feb 17b M		38. 3	23. 3	2. 0	2. 0	0. 06	0
- 462 Apr 30 M		38. 3	23. 3	2. 0	2. 0	0. 06	0
- 430 Aug 3a M		39. 4	24. 2	0	a	0. 02	1
- 430 Aug 3b M		38. 0	23. 7	0	0	0. 06	0. 1
- 393 Aug 14 M		38. 3	23. 3	0. 2	0.2	0. ?	1
- 363 Jul 13 M		38. 3	23. 3	0	0	0. 1	0. 1
- 309 Aug 15a M		37. 5	15. 2	0. 5	0	0. 01	0. 1
- 309 Aug 15b M		40. 2	26. 4	0. 5	0. 5	0	0. 24
- 281 Aug 6 M		40. 2	26. 4	0. 5	0. 5	0	0. 24
- 216 Feb 11 M		40. 2	26. 4	0. 5	0. 5	0	0. 24
- 189 Mar 14 M		40. 2	26. 4	0. 5	0. 5	0	0. 24
- 128 Nov 20 M		40. 2	26. 4	0. 5	0. 5	0	0. 03
- 124 Sep 7 M		40. 2	26. 4	0. 5	0. 5	0	0. 01
59 Apr 30 M	16 1/2	40. 0	44. 0	1. 2	1. 2	0. 1	1
360 Aug 28 M	09	37. 4	41. 8	2. 0	2. 0	0. 06	0. 5
418 Jul 19 M	14	40. 4	32. 8	0	a	0	1
484 Jan 14 M	~07	38. 0	23. 7	0	0	0. 01	0. 5
590 Oct 4 M		41. 0	29. 0	0	0	0. 06	1

[a]Cannot be assigned until path direction is known. See Section III. 1.

The passage is by the Greek poet Archilochus. Ginzel gives an independent translation that agrees with Fotheringham as far as the purposes of this study are concerned, and it does not seem necessary to find the original source.

High confidence has been placed upon the reliability of this report and upon the identification of the eclipse. There is little in the references, which between them summarize the results of all independent studies of the eclipse that I have found, to justify confidence in either the reliability or the identification.

This is a literary eclipse. If treated routinely it would receive a reliability of zero (see Section III.2). A different value can be justified only if one believes that there was little chance for Archilochus to come across a previous eclipse description in his reading, correspondence, or conversation. The horizon of a Greek poet in the -7th century was fairly large. The high cultures of Crete and Mycenae had come and gone, the Trojan War was far in the past, and Homer and Hesiod were already standard works, whether they had been written down or not. Archilochus [Ginzel, 1899] probably traveled as far as Sicily. I see no objective way to assess the probability that Archilochus had read or heard about an eclipse, but it must be fairly high. Further, even if it is assumed that the passage was the result of an observation, there is no assurance that it was an observation of a large solar eclipse. Nothing keeps it from referring to, for example, a "dark day".

After considering the possibilities, I arbitrarily assign a reliability of 0.06 to the passage when it is considered as an eclipse report.

If the passage does refer to an eclipse, there is no clue to the magnitude except that it was large. I shall take the value 0.06 for the standard deviation of the magnitude.

Archilochus spent his early life on the Aegean island of Paros, his later life on the island of Thasos, and an unknown amount of time at Sparta and in Sicily. I shall take the possible places of the observation to be Paros and Thasos. It makes negligible error to use the towns of the same names. Paros is at 37°.1N, 25°.1E and Thasos is at 40°.8N, 24°.7E. Strictly, the reliability assigned earlier should be multiplied by the fraction of his time that Archilochus spent at one of these places, but this refinement is unnecessary. Since the two possible places are nearly on the same meridian, it is simpler to use a region with 0° standard deviation in longitude and 1°.1 in latitude than to use the ordinary procedure for two points.

The record gives a fine chance to play the identification game of Section III.2. Since the dates of Archilochus are uncertain by several decades at each end, and since he must be allowed an active life of several decades as an observer, the game may be played over a century.

Oppolzer [1882] tentatively identified Archilochus' eclipse as that of -647. He cautioned that his studies about the identification made no claim to completeness. Fotheringham [1920] accepted his identification and ignored his caution. Ginzel [1899] calculated the circumstances of the five eclipses listed above for Paros and Thasos. He eliminated the eclipse of -661 on the basis that it was too small, but gave no details. He eliminated the eclipses of -656 and -660 on the basis that the calculated magnitudes lay between 0.90 and 0.96! He concluded that the choice between -688 and -647 had to rest upon external evidence about Archilochus' dates, and finally settled upon -647 as more probable on this basis.

There have been too many changes in the constants used in the ephemerides since Ginzel's work to let us accept the identification without recomputation. Further, I do not accept the criteria for rejection as being valid. Therefore I shall calculate this eclipse for all five dates listed above, giving each the same reliability. -688 perhaps pushes the historical possibilities a little, but the other four are well within accepted limits. I see no a priori reason to exclude the eclipse of -634 Feb 12 and shall also use it. This gives each date a reliability of 0.01.

-634 Feb 12 M.　See -647 Apr 6 M.
-607 Feb 13 M.　See -584 May 28 M.
-587 Jul 29 M.　See -584 May 28 M.
-584 May 28 M (Thales). Reference: [Herodotus, ca -446, Bk. 1]. The translation mentioned in the bibliography reads: ". . war broke out between the Lydians and the Medes, and continued for five years, with various success. In the course of it the Medes gained many victories over the Lydians, and the Lydians also gained many victories over the Medes.... As, however, the balance had not inclined in favor of either nation, another combat took place in the sixth year, in the course of which, just as the battle was growing warm, day was on a sudden changed into night. This event (eclipse)[†] had been foretold by Thales, the Milesian, who forewarned the Ionians of it, fixing for it the very year in which it actually took place. "

It does not seem that Thales' prediction was highly detailed. If he did predict the year, he probably did so by means of cycles[‡]. This is not the place to go seriously

---

[†]The translator interpolated this word. It (εκλειψις) does not occur in the original.

[‡]The cycle or interval of 223 synodic months, about $18^y 11^d$, was perhaps the first eclipse cycle discovered. It is remarkably successful for lunar eclipses, and also for solar eclipses if the entire earth is considered. It is not at all successful for solar eclipses that are large at a given point. It is common in 20th century literature to say that this cycle is the interval that the Babylonians or other ancients called the Saros, and it is probably hopeless to try to correct this error. Sarton [1952, p. 119] and Neugebauer [1957, p. 142] point out that the mistake arose in the 17th century. Halley started it by misreading a poorly edited text of Pliny's Natural History.

However, if there were a prediction by cycles, it is hard to understand the remark about "the very year". A cycle contains an integral number of months, and a prediction by cycles necessarily predicts "the very day". Perhaps Herodotus' source said "day" and Herodotus found such precision incredible.

into the question of whether Thales did predict this eclipse, but a remark about a commonly used argument may be useful.

This argument is that Thales did not make the prediction because cycles do not predict eclipses visible at a particular place with reliability. It seems to me that this argument misses the point of the legend. A legend that Thales predicted this eclipse, whether true or false, could not have existed if solar eclipse prediction had been reliable at the time. Because of their total reliability the eclipse predictions made in a modern ephemeris will never find a place in legend. Unreliable prediction was possible at this time, and the eclipse of Thales may commemorate one of the relatively rare successful predictions. †

This is one of the most discussed of eclipse reports. It should be noted that the word for eclipse does not occur, and that the description is compatible with other phenomena. However, the reference to Thales' prediction makes no sense unless Herodotus meant to describe a solar eclipse. Other aspects of the reliability will be discussed in connection with the record -477 Feb 17a M. If it be interpreted as an eclipse, the standard deviation of the magnitude would be taken as 0.06 on the basis of the description. However, if we accept the story at all, we must accept that the eclipse was predicted and would have been looked for. Thus a large standard deviation in magnitude must be taken, say 0.3.

With regard to the place, Fotheringham [1920] says: "It has frequently been assumed that the battle described by Herodotus must have been in the neighbourhood of the Halys, presumably because Croesus crossed the Halys when he made war on Persia. But this is quite fanciful. The battle may have been fought almost anywhere in Asia

---

†Neugebauer [1957, p. 142] says that "there exists no cycle for solar eclipses visible at a given place; . ." This is correct for long series of eclipses, but cycles will yield pairs and occasionally triads of visible eclipses. Cycles for solar eclipses work just often enough to be tantalizing.

Minor." Fotheringham overlooks many other clues to the location of the battle that Herodotus gives, and it is highly plausible that the battle took place near the Halys, a river in central Asia Minor. Fotheringham [1921] himself a year later leaned toward a location in central Asia Minor but slightly south of the Halys. However, for reasons that will appear later and that have nothing to do with the place of the battle, I shall accept anywhere in the western half of Asia Minor (37° to 41°N, 27° to 35°E).

The only clues to dating the battle are the names of the kings involved. According to the unsigned articles on these kings in the Encyclopedia Britannica, the reign of the Lydian king Alyattes was from ca -608 to ca -559 and the reign of the Median king Cyaxares was from ca -623 to ca -583. Webster's Biographical Dictionary [1953] gives ca -616 to ca -559 for Alyattes and ca -624 to ca -584 for Cyaxares. It is not clear what part the identification of the eclipse as that of -584 has played in these dates, but it has almost surely played some. Taking the dates as given, with a cushion of 5 years at each end, we have from, say, -613 to -578 to look for the eclipse. The name Thales does not help, because he can be dated only by means of the eclipse.

The time interval should perhaps be extended on the near side. Fotheringham [1920] mentions several ancient writers besides Herodotus who referred to the eclipse of Thales. Three of these, about half of the total who mention the eclipse, put the events in the reign of Cyaxares' son Astyages who reigned for about 35 years. This would allow basing the near end of the search interval on the death of Alyattes in ca -559. However, there were not many eclipses in the right part of the world for some time after -578. The next one that looks likely on the basis of Oppolzer's Canon is that of -556 May 19. Since this is three years after the traditional date of death of Astyages, I shall ignore it, perhaps wrongly.

The three eclipse dates listed above, and -581 Sep 21 given below, look reasonable on the basis of the charts in Oppolzer's Canon. Still other choices can be defended, and Ginzel [1899] lists about 15 eclipses that have been

96

suggested and studied. I shall, with reservations, stay with four, giving one fourth of the report reliability to each.

-581 Sep 21 M. See -584 May 28 M.

-556 May 19 M (Larissa). References: Ginzel [1899], Newcomb [1875]. Both refer to a passage by Xenophon, Anabasis III, 4. 8. Newcomb states that the late Astronomer Royal Airy translated the passage (in an uncited work) as follows: "When the Persians obtained the empire from the Medes, the king of the Persians besieged this city, but could not in any way take it. But a cloud covered the sun and caused it to disappear completely. . . ." Ginzel's translation is in accord. It seems unnecessary to try to find the primary reference.

Newcomb also quotes Airy as saying: It cannot be doubted, I think, that the disappearance of the sun at Larissa was caused by a total eclipse. " Newcomb says that he cannot "share the confidence of the Astronomer Royal. " Nor can I.

The darkening agent in Xenophon is clearly identified by the word νεφέλη or "cloud". If there were independent evidence that dated the siege closely, and if this date should coincide closely with an eclipse, we might guess that Xenophon had erred, and equate the "cloud" with the moon. However, it seems that the siege can be dated only by the "eclipse". Only romance could call this a useable eclipse record, and I shall not calculate it.

-487 Sep 1 M. See -462 Apr 30 M.

-477 Feb 17a M. Reference: Herodotus [ca -446, Bk. 7]. In the cited translation, this reads: ". . . then at length the host, having first wintered at Sardis, began its march toward Abydos, fully equipped, on the first approach of spring. At the moment of departure, the sun suddenly quitted his seat in the heavens, and disappeared, though there were no clouds in sight, but the sky was clear and serene. Day was thus turned into night; . . ." Xerxes sent for the Magians who said ". . the sun foretells for them, and the moon for us. " The Magians, wrongly as it turned out, told Xerxes that the portent was favorable and Xerxes proceeded "with great gladness of heart". There is little question that Herodotus is talking about a large eclipse of the sun here, even though the word "eclipse" is not used.

In Book 9, Herodotus describes the preparations of the Greeks to meet the threatened invasion by Xerxes. A Lacedaemonian force went to fortify the Isthmus of Corinth and returned. Its leader Cleombrotus "died a short time after bringing back .. the troops who had been employed in building the wall. A prodigy had caused him to bring his army home; for while he was offering sacrifice to know if he should march out against the Persian, the sun was suddenly darkened in mid sky."

As with Sherlock Holmes' curious incident of the dog who did not bark in the night, the curious point about the Sardis eclipse is that it did not happen. Xerxes' departure from Sardis can be dated accurately from historical records. I have never seen any date seriously suggested except -479 or -480, and the standard date is -479. No eclipse of magnitude enough to justify Herodotus' description was visible in Asia Minor in the early part of either year.

Herodotus wrote his history after spending some time studying and interviewing people in all the areas of importance in the Xerxes campaign. He did this within a generation of the events and probably talked to many participants in them. Thus the eclipse at Sardis is perhaps an assimilated eclipse; if so, identification with the eclipse of -477 Feb 17 is the most probable one. This eclipse has the advantage of occurring fairly close to spring. Rather arbitrarily, I give this report and identification a reliability of 0.1. If the eclipse is assimilated, the place of observation could be at a considerable distance from Sardis ($38°.5N$, $28°.0E$). I adopt the location with a standard deviation of $1°$ in both latitude and longitude. I also give it a standard deviation of 0.06 in magnitude.

The eclipse at the Isthmus of Corinth is probably assimilated also. Since the description does not make it sound large, it does not seem either identifiable or usable on the basis of present knowledge.

Herodotus' lack of accuracy with regard to the eclipse at Sardis reacts upon the reliability of the eclipse of Thales. If he could err so badly within a decade or so, how badly could he err after more than a century? I shall assume, with a reliability arbitrarily taken to be 0.04, that there was a tradition in western Asia Minor in Herodotus' time of an eclipse associated with Thales, and that this eclipse was as-similated[†] to the undatable battle between the Medes and the Lydians. The location of the battle thus becomes irrelevant. Since there are four identifications of the eclipse, each re-ceives a reliability of 0.01.

Only Newcomb [1875], of the studies I have read, considers the eclipse at Sardis while assessing the eclipse of Thales. The usual tendency is to take Herodotus' de-scription of Thales' eclipse, which cannot be tested, at more than face value and to ignore Herodotus' other eclipse reports, which can be tested and shown false.

-477 Feb 17b M. See -462 Apr 30 M.

-462 Apr 30 M (Pindar). Reference: Fotheringham [1920]. Fotheringham gives translations of two passages by the poet Pindar: "Beam of the Sun! O thou that seest afar, what wilt thou be devising? O mother of mine eyes! O star supreme, reft from us (κλεπτόμενον) in the daytime! Why hast thou perplexed the power of man and the way of wisdom, by rushing forth on a darksome track?" And: "God can cause unsullied light to spring out of black night. He can also shroud in a dark cloud of gloom the pure light of day." Fotheringham remarks that κλεπτόμενον as used is consistent with either complete or incomplete action.

If this be an eclipse, it is surely a literary eclipse. We are now two centuries after Archilochus, and we are one century after the time when Thales was traditionally predicting eclipses. Since Plutarch (see the report 83

[†]Some of the ancient writers whom Fotheringham [1920] mentioned give the battle and the eclipse as different events. It seems likely that there were two independent traditions of the eclipse in the ancient world.

Dec 27 M), five centuries later, knew Archilochus' eclipse description, Pindar certainly had the opportunity to know it. I see no reason to infer from the passages that Pindar had seen a large eclipse of the sun. If the passages were suggested to Pindar by the sight of an eclipse, or the hearing of an eclipse prediction, there is no reason to assume that the eclipse was a large one. I give this record a reliability of zero.

If the reader wishes to assume that this is a report of a large solar eclipse, he still has the problem of identification. The standard dates of Pindar [Webster's Biog. Dict., 1953] are ca -521 to -442. We are free to play the identification game from, say, -500 to -450 for the region of classical Greece. Three large eclipses during this interval are those of -487, -477, and -462. Fotheringham [1920] identified the eclipse as -462. The sole basis of this identification is apparently a calculation made in 1821 that this eclipse had the largest magnitude at Thebes of those allowed by the time interval. Use of Thebes in this calculation is based upon a tradition that Pindar was born near there; he apparently lived in several places.

Still on the assumption that this is a large solar eclipse, the standard deviation of magnitude should be taken as 0.06. The place should probably be taken as Thebes ($38°.3N$, $23°.3E$) with standard deviations of $2°$ in latitude and longitude. This allows observation anywhere in ancient Greece.

-430 Aug 3a M (Thucydides). Reference: Thucydides [ca -420]. The translation mentioned under this reference reads: "The same summer, at the beginning of a new lunar month, the only time by the way at which it appears possible, the sun was eclipsed after noon. After it had assumed the form of a crescent and some of the stars had come out, it returned to its natural shape."

Thucydides was writing a history of the Peloponnesian wars. He played a considerable part in the politics of their time and wrote almost contemporaneously with the events he describes. He made no attempt to turn the eclipse into a marvel or to associate it with any great event. The reliability of this record will be taken as unity. The record can be dated by historical evidence and there is apparently no question about the identification[†]. Since he explicitly denies totality while mentioning the visibility of stars, I shall take the standard deviation of magnitude to be 0.02 rather than the customary value of 0.01 when stars are mentioned. It remains to discuss the place.

Thucydides was an Athenian. Most students of this eclipse have felt compelled to take Athens as the place and have struggled with the fact that their computations place the zone of annularity considerably to the east. Some have tried to find political or diplomatic reasons to send Thucydides to eastern Greece that summer. There is no need for a complex reason; there is a simple homely reason. According to Jebb [1911], Thucydides' family derived its wealth from the ownership of gold mines on the mainland coast opposite the island of Thasos. The ownership of important property here must have required frequent visits and frequent correspondence. It is almost certain that an eclipse seen at the mines would have been reported to him whether he were there or not. Thasos, which is close to the coast, is thus one possible place of observation, and Athens ($38^{\circ}.0N$, $23^{\circ}.7E$) is an equally probably alternate. It is a reasonable speculation that Thucydides had and combined reports of the eclipse from both places. This would account for the possible contradiction in his report between the crescent shape and the visibility of stars.

-430 Aug 3b M. Reference: Plutarch [ca 100, Life of Pericles]. On page 72 of the translation mentioned in the citation we find: ". . . and Pericles being gone aboard his own galley, it happened that the sun was eclipsed, and it grew dark on a sudden, to the affright of all, for this was

---

[†]Some have tried to identify this record with the eclipse of -433 Oct 4, but apparently this is outside the limits allowed by the historical evidence. Further, October is a bit late to call summer, although not impossibly so.

looked upon as extremely ominous. " The incident took place on an Athenian ship preparing to leave for a raid on the Peloponnesus.

Independent evidence dates the raid described in the summer of -429 rather than -430. I do not know whether the historical evidence allows an uncertainty of a year or not. In any case, there was also an Athenian raid upon the Peloponnesus the preceding summer. Plutarch could have made a simple error in the raid that he connected with the eclipse, and I see no strong reason to question the identification[†].

Plutarch wrote about five centuries after the event, using Thucydides and other sources. The eclipse report came from a source that we know little about. This could be an assimilated or a literary[‡] eclipse. Little is made of the marvelous, so it is probably not a magical eclipse. Since the event is in the right season and only a year from a known eclipse, I shall give the report a reliability of 0. 1. The standard deviation of the magnitude will be taken as 0. 06. The place will be taken as Athens, although strictly speaking one should use the port of Piraeus a few kilometers away.

-393 Aug 14 M. Reference: Ginzel [1899]. Ginzel quotes a passage from the Greek historian Xenophon, Hellenica IV, 3. 10. A translation of Ginzel's German is: "As he (Agesilaos) was invading the country (of Boeotia), it happened that the sun took on the form of a crescent. " With this straightforward text, it seems safe to rely upon the translation of a translation.

[†]If -433 Oct 4 should become allowed for the eclipse of Thucydides, it should also be allowed here. October is late for the start of a raid, but the discrepancy could be the result of assimilation.

[‡]Since Pericles [Plutarch, ca 100] used the eclipse to teach his sailors a lesson, we may need a category called the "moral eclipse".

102

This was contemporary history when it was written. It does not seem that any attempt was made to make the eclipse into a marvel, and we can give the report unit reliability. At most a small amount of assimilation might have occurred. The year can be found from independent evidence. Since we are now at a period in Greek history when systematic astronomical observations were being made, I increase the standard deviation of the magnitude to 0.1. I assign the place to be Boeotia. The position of Thebes, with standard deviations of $0°.2$ in latitude and longitude, will do.

-363 Jul 13 M. Reference: Ginzel [1899]. Ginzel quotes passages from Plutarch (Pelopidae, 31) and from the historian Diodorus (XV; 80) that mention an eclipse in Thebes in -363 or -364. Since both writers are several centuries after the event, I give the report a reliability of 0.1. The identification seems certain. 0.1 will be used for the standard deviation of magnitude. The place will be taken as Thebes.

Ginzel also refers to an eclipse observation in Syracuse that is mentioned by Plutarch (Dionis, 19). The context does not allow dating the eclipse securely. Further, the observation was made in order to test a prediction and we can infer nothing about the magnitude. I shall not use this record.

-309 Aug 15a M (Agathocles). Reference: Diodorus, [ca -20]. The translation mentioned in the references reads: "On the next day there occurred such an eclipse of the sun that utter darkness set in and the stars were seen everywhere; . . ." The observation was made from a ship of the Syracusan tyrant Agathocles as it was desperately and successfully running a Carthaginian blockade. The observation was made the day after the ship left Syracuse, apparently in late afternoon, on a voyage of six days to a point near Carthage.

There is no way to tell whether the ship went north or south from Syracuse in clearing Sicily. Fotheringham

[1920] took the possible points of observation to be 38°.4N, 15°.5E and 36°.6N, 15°.0E, and these points seem reasonable. Curott [1966] used 37°N, 13°E. This is about halfway to Carthage and seems too far west. Ginzel [1899] used slightly different points from Fotheringham, but the difference is not important. It is convenient, considering the locations of the possible places, to use the mean position with zero standard deviation in longitude and 0°.5 in latitude.

The history is quite detailed and the dating of the eclipse seems secure.

The translation quoted above says that "utter darkness set in"; this sounds like totality. Fotheringham quotes from a translation dated 1700 that I do not have; it reads instead that "day was turned into night". The Greek has ὥστε ολοσχερως φανηναι νυκτα. This perhaps means "as appearing completely like night". It is hard to decide what degree of totality this implies, but absolute totality cannot be assumed safely. Since stars appeared, I give the value 0.01 to the standard deviation of the magnitude.

The eclipse occurred during a dramatic moment and it was interpreted by the participants as a bad omen. There is strong suspicion that the eclipse is magical. Diodorus did not stress the magical aspect; he recounted the eclipse and the interpretation as a chronicler would. However he wrote three centuries after the voyage and we do not know how his sources treated their material. There certainly was an eclipse at this time in this general part of the world, but the sources could have assimilated an eclipse from some other point to the blockade running that was part of a major war. I shall give the report a reliability of 0.1. The low value indicates doubt, not about the fact of the eclipse, but about the accurate reporting of the place.

-309 Aug 15b M (Hipparchus). Reference: Fotheringham [1920]. Fotheringham translates a passage by the 1st or 2nd century astronomer Cleomedes (De Motu

<u>Circulari Corporum Caelestium</u>, <u>2</u>, 3) thus: "Once it[†] was observed to be wholly eclipsed at the Hellespont, when at Alexandria it was eclipsed with the exception of one-fifth of its own diameter, that is, in appearance with the exception of two digits[‡] and a fraction." Fotheringham also translates a passage by the astronomer and mathematician Pappus (<u>Commentary on the Fifth Book of the Almagest</u>, ca 300): "For in his first book concerning magnitudes and distances he[*] uses the following phenomenon: - an eclipse of the Sun which was actually of the whole Sun in the places round the Hellespont, so that no part of it could be seen at the Moon's edge, while at Alexandria in Egypt the Sun was eclipsed to the extent of about four fifths of his diameter." The Greek texts of both passages are given by <u>Fotheringham</u> [1909a]. Since the passages are from technical works by writers who still had access to Hipparchus' writing and to many of his records, it does not seem necessary to go back to the primary references. <u>Ptolemy</u> [ca 152, V. 11] alludes to Hipparchus' use of this eclipse but gives no details of the observations. Hipparchus used the records of this eclipse to calculate the lunar parallax.

Cleomedes, Pappus, and Ptolemy all refrain from identifying the eclipse. We can try to identify it only by combining the man Hipparchus, the places Hellespont and Alexandria, the magnitudes, and the results of calculation. The time limits for the identification are given by the dates of Alexandria and Hipparchus. Alexandria was founded in -331. The last known observation by Hipparchus was on -127 Mar 23. He probably died or retired soon after, but to be safe we should carry the end date forward to, say, -120. We have more than two centuries for playing the identification game.

[†] That is, the sun.

[‡] The digit means a twelfth part. See Section V.1 for further discussion of this measurement.

[*] The context refers to Hipparchus.

In recent literature, identification with the eclipse of -128 Nov 20 has become standard,[†] apparently on the authority of Fotheringham. It is fascinating to see how this identification became adopted.

Fotheringham [1920] started by considering only the eclipses of -309 and -128, on the basis that all other possibilities had been eliminated by earlier authorities. He conscientiously carried out many of the calculations for -309 but did not draw the lines for it on his graph.[‡] He preferred the eclipse of -128 for two main reasons. These I quote below and follow with my comments.

First: "There is also the inherent probability that Hipparchus would collect two observations of an eclipse in his own time more easily than two observations of an eclipse 180 years earlier."

Astronomers must use the data that they have; they are not at liberty to perform experiments. Hipparchus freely used data from earlier times, of which he apparently possessed large amounts. It is on record [Ptolemy, ca 152, III. 2] that Hipparchus considered an observation made in Athens in -431 and rejected it because he did not think it was accurate enough for his purpose. He used a number of observations between -431 and his own time. Weather permitting, the eclipse of -309 was probably seen near the Hellespont and in Alexandria. If so, Hipparchus probably had access to the records. The dates of the Alexandrian observations in Section VI. 2 should also be noted.

If there is any inherent probability connected with the dates of -309 and -128, it favors the earlier date. Hipparchus stopped work a year and a half after the eclipse of -128. He had a professional life of at least 34 years.

[†] Although van der Waerden [1961] tried to warn against it.

[‡] This refers to the much-reproduced graph that Fotheringham used in making his final inferences. I have verified that at least one of the "lines" for the eclipse of -309 fits nicely on his graph.

The a priori odds are thus about 20 to 1 ($\approx 34/1.5$) that Hipparchus made his famous calculation before the eclipse of -128 and hence that his eclipse was not that of -128.

Second: "I doubt, also, whether an astronomer of -309 would be likely to think of estimating, much less of recording, the magnitude of a solar eclipse. The earliest Greek observation of a lunar eclipse, cited by Ptolemy from Hipparchus, is that of a partial eclipse at Alexandria -200 September 22, where the magnitude is not recorded. From that date to -140 four more lunar eclipses are cited, of which two, in -173 and -140, were partial. In both cases the magnitude is recorded."

Fotheringham's statement about the recorded magnitudes is literally correct, but the implication comes from his emphasis, not from the records. The records cited are from a series [Ptolemy, ca 152, IV.5] of 19 lunar eclipses. Four are from Ptolemy's time; the remaining 15 run from -140 (in Hipparchus' time) back to -720. The ones before -200 are from Babylon. Four of the early ones were total and were so recorded. Of the 11 still remaining, the magnitude is recorded for eight, including six from -490 back to -719. Of the three eclipses with no recorded magnitude, two were in progress at moonrise or moonset. It is not sure that this is the reason for not recording the magnitude. The three partial eclipses for which the magnitude was not recorded are the three that occur in Chapter IV.10, where Ptolemy was interested in the longitudes of the sun and moon during the eclipse and not in the magnitude. The magnitude is recorded for every partial eclipse used in every other part of Ptolemy's work.

Thus it is correct to say that the magnitude (including totality when appropriate) was recorded in every known lunar eclipse record from -720 through +136, with one exception, when the eclipse could be observed in all its phases. In the one exception, the magnitude was not relevant in the context where it was preserved, and it is probable that the magnitude was originally recorded for every partial eclipse. This paragraph, which is correct on the basis of the record, has quite different implications from Fotheringham's literally correct statement.

Cultural contacts between the Greeks and Babylonia-Assyria go back to an unknown date. They quickened after the Persian Wars[†] around -475 and became close after the campaigns of Alexander and the founding of Alexandria, more than 20 years before -309. Around this time the Greeks were at a high level of intellectual vigor and curiosity. Thus it is highly probable that the Greeks learned about eclipse magnitudes soon after -331 if they did not already know.

As I have said, Fotheringham [1920] started by eliminating all possible identifications except -309 and -128. It would be too lengthy to give the basis for this in detail. He depended upon eclipse magnitude calculations made by other astronomers before 1900 with already obsolete astronomical constants. He used an additional argument [Fotheringham, 1909b] that will be discussed in Section V. 1. It will be shown there that the conclusion given by Fotheringham [1909b] does not follow from his premises and further that Fotheringham [1920] himself disproved a major premise of his 1909 argument without noticing the connection between the premise and the identification. It is appropriate to reopen the question of identification. [‡]

The following identifications are reasonable on the basis of the charts in Oppolzer's and Ginzel's Canons: -323 May 23 (ann), -309 Aug 15, -281 Aug 6, -262 Feb 9

[†] Herodotus could travel as far as Persia to collect material.

[‡] Neugebauer [1929] did so and settled on -309 Aug 15. This conclusion was reached by his own version of the identification game and is open to as much objection as Fotheringham's identification.

(ann), -251 Jul 7 (ann), -216 Feb 11, -208 Mar 13 (ann),
-189 Mar 14, -128 Nov 20, and -124 Sep 7 (ann. tot.). The
eclipses followed by the notation (ann) were annular and
that of -124 was annular-total. Mostly to save (probably
useless) work and partly because the observers may have
known to distinguish a total from an annular eclipse[†], I
shall drop the purely annular eclipses. This leaves six
possibilities.

I give the record a reliability of 1, distributed among
the six identifications retained, as follows: 0.24 for those
through -189 (before Hipparchus' time), 0.03 for -128, and
0.01 for -124. This is a reasonable a priori distribution
of weights, made on the record alone before detailed calculation. It assigns a lower weight as we approach the end of
Hipparchus' work.

The standard deviation of the magnitude will be
taken as zero for the observation at the Hellespont. The
use of the observation at Alexandria will be discussed in
Section V. 1. The standard deviation of position for the
Hellespont observation needs discussion.

The passage from Cleomedes is less detailed in all
respects than the one from Pappus and does not help us decide the exact place[‡]. Pappus uses "at the places near the

---

[†] It is probable that the distinction was not known even in
Ptolemy's time. Ptolemy [ca 152, VI. 7], in his eclipse
tables, used a constant angular diameter of the sun, and
he stated that the minimum diameter of the moon is the same
as the diameter of the sun. This statement seems incompatible with a knowledge of annular eclipses.

[‡] It is unlikely that Pappus would supply detail that did not
exist in the record available to him. The usual tendency
in a technical report is to drop detail at the expense of accuracy.

Hellespont" (εν μεν τοις περι τον Ελλησποντον), and
Fotheringham [1909a] comments on the vagueness of this
phrase. In spite of this vagueness, Fotheringham chose to
take the place exactly on the Hellespont, on the basis that
Hipparchus did so in his calculation. But Hipparchus had a
different interest. He needed the distance from Alexandria
to the path of totality, and an error in longitude at the
Hellespont did not matter much for this purpose. For us,
an error in longitude is more important.

It is hard to know how far away is "near". It is possible that the observation was made from one of the Greek
cities on the Asian coast of the Aegean, for example, and
that "near the Hellespont" meant that it was closer to the
Hellespont than to the open Mediterranean. I shall not push
the term that far. I shall take the central point to be the
town of Dardanelles, the present Çanakkale (40°. 2N, 26°. 4E),
at the narrowest point of the strait, and use standard deviations of 0°. 5 in latitude and longitude.

> -281 Aug 6 M. See -309 Aug 15b M.
> -216 Feb 11 M. See -309 Aug 15b M.
> -189 Mar 14 M. See -309 Aug 15b M.
> -128 Nov 20 M. See -309 Aug 15b M.
> -124 Sep 7 M. See -309 Aug 15b M.
> +29 Nov 24 M (Phlegon). Reference: Eusebius

[ca 325]. Eusebius was a famous church figure who took a
leading part in the Council of Nicaea in 325. One of his
many writings was the Chronicon, a chronicle of events
from the time of Abraham to his own time. The original has
been lost but there are several replacements. There is a
translation into Armenian (which was in turn translated into Latin in the edition cited in the references) made in the
5th century. There is a translation into Latin by St. Jerome
(Eusebius Sophronius Hieronymus) which, along with a continuation of the chronicle, was prepared about 380. There
are considerable differences between the two versions, and
it looks as if at least one of them was a use of Eusebius'
Chronicon in preparing a new work rather than a mere
translation. Fortunately, the versions agree closely in the
part of immediate interest, and we can safely assume for
present purposes that we have a text close to the one that

Eusebius wrote. Portions of the Chronicon are preserved in a text by Syncellus written about 800 and in a Syriac epitome made about 635, but these have less evidential value.

The passage of interest says approximately: "Jesus Christ, the Son of God, in accordance with the prophecies made about Him, went to His Passion in the 19th year of Tiberius; for this time we find indeed in different Greek records the following related word for word: the sun was extinguished, in Bithynia an earthquake happened, which overturned the greater part of Nicaea, -- that agrees with what happened at the Passion of our Lord. Thus Phlegon, who wrote the Olympiads, reports the following word for word in his thirteenth book: 'In the fourth year of the 202nd Olympiad, there was an eclipse of the sun which was greater[†] than any known before and in the sixth hour of the day it became night, so that stars appeared in the heaven; and a great earthquake that broke out in Bithynia destroyed the greatest part of Nicaea.'" Fotheringham's [1920] translation of the "Phlegon" part of the passage agrees well with this. If Phlegon wrote this word for word, he was badly informed.

As it stands, this is a magical eclipse. It has to be magical if for no other reason than an unstated one: because of the way Passover is determined, the Crucifixion had to occur within a day or so of the full moon. Assimilation may underlie the magic. If we had Phlegon's original writing available, we might be able to judge. Since two of Phlegon's works were called On Marvels and On Long-Lived Persons (about Italians who lived more than 100 years), we are entitled to be suspicious about his credulousness and his credibility. Since he wrote a century after the event, from unknown sources, and since all we have is a paraphrase made several centuries later yet by a writer with a point to prove, I see no reason to attach any reliability to this record.

[†]The wording of the Syriac epitome suggests that we should use "longer" rather than "greater".

Phlegon's Olympiads were a compendium of history from about -775 to +140. Presumably he used events known to him that happened during this time in the world of his time, that is, roughly the Roman Empire[†] at its greatest extent. I do not see any reliable way to narrow the place of observation, if indeed there were an observation. There is a mild presumption in favor of the eastern part of the Roman Empire since Phlegon was born in Asia Minor and spent most of his life in the eastern part of the Empire.

If we assume that Phlegon's eclipse is to be explained by assimilation and not by pure magic, we must still identify it. The year stated in two different ways formed part of our years 32 and 33. By Oppolzer's Canon, the closest eclipse to any part of the Roman Empire during either year was the eclipse of 33 Sep 12, which passed over the North Pole and through central Siberia; it is doubtful that this was Phlegon's eclipse. In a case of ordinary assimilation, we would choose the eclipse closest in time that was total near the right part of the world; this was the eclipse of 29 Nov 24.

This is not a case of ordinary assimilation. Usually we assume, when assimilation has probably occurred, that an event and an eclipse have been assimilated to each other by people who knew each in some way and who confused the dates. In such a case, every eclipse is a possibility and it is plausible to choose the closest in place and time. In this case, assimilation, if it occurred, was done by a writer several centuries later who could only go by recorded eclipses and not necessarily by all observed eclipses. Since the writer (Eusebius) apparently did not use technical records, if we may judge by the wording, but instead used now lost non-technical records, we have no idea what eclipses were available to him for assimilation. Hence any identification is a weak assumption.

[†]An educated Roman of this time had some knowledge of India, China, and other distant places, but probably not historical knowledge of them.

112

Since this eclipse record has negligible reliability, weak identification, and uncertainty of position within most of the Roman Empire, it does not seem worthwhile to calculate it.

Several recent papers place this eclipse observation at Nicaea or in Bithynia and cite Fotheringham [1920] as the authority for doing so. Fotheringham did not place the observation so confidently[†]. He said: "Since the zone of totality must have passed through or near Bithynia, it seems reasonable to presume that it was the author's intention to place both the totality and the earthquake at Nicaea, and I have accordingly assumed that the eclipse was total at Nicaea. But I consider the presumption made in the case of this eclipse weaker than any of the other presumptions used in this paper." Recent writers have overlooked the element of strong doubt. Thus the known world has shrunk to a point.

It is doubtful that Eusebius' intention was dependent upon knowledge of a calculation of the path of totality that would be made 16 centuries later. It should be noted that he twice rather carefully separated the eclipse[‡] from the events in Bithynia and Nicaea. Considering his purpose in writing, it is reasonable to assume that he in fact meant to make the eclipse world wide; we do not know what Phlegon intended.

59 Apr 30 M. Reference: Ginzel [1899]. The eclipse was seen in Italy and Armenia; the report is discussed under

---

[†] Nor did Ginzel [1899], who said that there is a good possibility but not certainty that the place was Nicaea, or at least Bithynia.

[‡] In the originals as well as in the translation given here. Further, Eusebius noted several other eclipses; in no case that I noticed did he locate the observation. He noted many other earthquakes and located all of them. In the relevant passage, it may be that he merely continued these habits with no particular intention in mind.

the listing 59 Apr 30 E. Armenia was approximately the region from 30° to 42°N and from 42° to 46°E.

71 Mar 20 M. See 83 Dec 27 M.
75 Jan 5 M. See 83 Dec 27 M.

83 Dec 27 M (Plutarch). Reference: <u>Plutarch</u> [ca 90; Chap. 19]. "Now, grant me that nothing that happens to the sun is so like its setting as a solar eclipse. You will if you call to mind this conjunction recently which, beginning just after noonday, made many stars shine out from many parts of the sky and tempered the air in the manner of twilight. If you do not recall it, Theon here will cite us Mimnermus and Cydias and Archilochus and Stesichorus besides and Pindar..." The passage goes on to quote descriptions of eclipses by the writers mentioned, as well as a possible eclipse description in Homer. Further on, the passage says that, during a solar eclipse, "a kind of light is visible about the rim which keeps the shadow from being profound and absolute". The quotations are from the translation mentioned in the references.

The "kind of light" may or may not[†] be the corona; <u>Dreyer</u> [1877] assumed that it was. If it be assumed that it is the corona, its status as the oldest mention of the corona depends upon the interpretation given to the report listed as -1062 Jul 31 BA (Babylon). If the latter be accepted as a description[‡] of a solar eclipse, it contains the oldest known possible mention of the corona (or prominences).

[†] Since the distinction between total and annular eclipses may well not have been known at this time, the "light" may instead be the rim of the solar disk that remains visible during an annular eclipse. See the discussion of the record listed as -309 Aug 15b M (Hipparchus).

[‡] The text could describe an eclipse without being a record of a specific or genuine eclipse.

The cited work has the form of a conversation. The speaker of the quoted part is called Lucius; "Theon here" is one of the participants.

Ginzel [1899] and Fotheringham [1920] discussed this eclipse report at great length. Fotheringham identified it as the eclipse of 71 observed at either Delphi or Chaeroneia (which are about 35 km apart), but he said (page 124) that the conclusions "fall short of certainty"[†]. Ginzel made the same identifications with more confidence: "This eclipse is suited to be a most important test of lunar theory, because of the narrow zone (of totality) that is described and of the complete certainty about its circumstances in time and place." I am like Newcomb confronted with the Astronomer Royal; I cannot share this confidence.

I am joined by other doubters. Cherniss [1957] cites some recent controversy on the subject. He leans toward the eclipse of 75 seen at Rome, and cites others who prefer the eclipse of 83 seen at Alexandria. Cherniss and those whom he cites apparently use the same approach as Fotheringham and Ginzel in reaching their conclusions. I have what seems to be a different kind of reason for doubt.

Attempts to identify the eclipse generally begin with an attempt to locate the place where the dialogue presumably occurred. With this goal, the writers dissect the written dialogue as if it were a careful transcript of a tape recording[‡]. The dissection is an interesting literary

---

[†] The qualification surrounding his conclusions has usually been omitted in quotations or citations from his paper.

[‡] Thus Fotheringham concluded that the conversation almost surely took place in Delphi or Chaeroneia. Cherniss is sure that it was in or near Rome.

115

exercise but does not help locate the time or place of the alleged eclipse observation. It is clear that the time of the "conversation" differs from the time of the eclipse. There is no indication that the places are the same, as Cherniss has pointed out[†]. There is also no reason to assume that Plutarch intended a particular site for the conversation.

The work is not a transcript of a tape recording of a conversation. It is a conscious literary invention by a writer who was skillful at inventing dialogue to illustrate his points. In order to judge whether Plutarch meant to describe a particular eclipse seen at a particular place, we must consider the purpose of his work and the purpose of the passage quoted, and we must consider his background.

The passage occurs in a "popular book" on astronomy. Plutarch was well equipped for the job of writing this book. He cited (Chap. 10) Hipparchus and Aristarchus[‡] among others. He showed a good understanding of the phenomena he described, within the limits of his time, and he argued (Chap. 6) that the moon does not fall because its "weight has its influence frustrated by the rotatory motion"; this point is not understood by many technically educated people today. A general description of a solar eclipse in a popular text on astronomy written for the "educated layman" does not create the slightest presumption in my mind that the writer has recently seen a total eclipse in person. An astronomy student who cannot describe an eclipse without seeing one will probably not pass the course.

[†]On the basis of several nuances in language, Fotheringham tried to argue that the conversation and the "eclipse" were close together, overlooking the possibility that Plutarch may have invented these nuances in order to give "verisimilitude to an otherwise bald and unconvincing narrative" [Gilbert, 1885].

[‡]Of Samos, who advocated a heliocentric theory and who estimated the distances to the sun and moon; his accuracy was limited by the observational methods available. He flourished around -250.

In particular the description creates no presumption of an eclipse actually seen when the writer immediately follows his own description by quoting eclipse descriptions from writers who were already classical; we have used two of these descriptions[†] in this paper. Plutarch shows us clearly that he was well acquainted with eclipses, both astronomical and literary.

Further, Plutarch needed an eclipse description at this particular point in the book in order to develop his point. He was arguing that the moon is a solid body, and not a self-luminous gas, for example. He demonstrated his point by remarking that the moon and earth both cast shadows, the one at twilight and the other during a solar eclipse, and that the effects of the shadows are similar. If he had not seen an eclipse, he would have had to invent one, as Clemens [1889] did. He demonstrably had the background needed to invent this particular eclipse.

This does not mean that Plutarch had not seen an eclipse; if he had, he did not tell us where or when and he had no reason to. His argument could have been suggested by seeing an eclipse, whether partial or total. The eclipses of 71, 75, and 83 were probably seen, weather permitting, by someone he knew somewhere. Considering his known interests, he probably heard of them wherever in the Roman Empire they were seen. Any eclipse description could have suggested the argument.

I do not take the passage from Plutarch to be a description of a specific eclipse. If it be one, it is unidentifiable both in time and place. I shall not calculate it.

218 Oct 7 M. Reference: Dio Cassius [ca 230, Bk. lxxvii, 30]. Dio says that there was a conspicuous eclipse during the time of the emperor Macrinus. Dio is not notable for his careful treatment of eclipses, but we are now in his own time and could expect higher reliability. Unfortunately, this part of Dio is preserved only in an epitome made in

---

[†]Archilochus and Pindar (-647 Apr 6 M and -462 Apr 30 M).

the 11th century. There was no large eclipse in any part of the Roman Empire during the short reign of this emperor (217 Apr to 218 Jun). The nearest in time is the eclipse of 218 Oct 7. Whether the error is due to Dio or to his epitomizer is unknown.

Dio was administrator of Pergamum and Smyrna in Asia Minor in the reign of Macrinus but probably returned to Rome after the successful revolution of Heliogabalus, which ended the reign. Even if we should assume that Dio saw the eclipse himself, we could not locate it. This eclipse does not seem worth calculating.

Ginzel [1899] will be the reference for the remaining reports from Mediterranean countries; they all come from histories or annals of the late Roman Empire and this careful secondary reference seems adequate for this class of report. Ginzel always reproduces the relevant quotations in the original language.

360 Aug 28 M. During a battle on the Tigris River near the point $37^{\circ}.4N$, $41^{\circ}.8E$ in the year 360 there was an eclipse of the sun, described by the contemporary historian Ammianus Marcellinus. According to him, "the stars remained visible from dawn until nearly noon." This eclipse occurred during a battle and smacks of the magical as well as the exaggerated. However, the year is given and the report is contemporary, so the eclipse is probably assimilated only a small amount in time and place. I shall give the report a reliability of 0.5 and a standard deviation of 0.06 for the magnitude, ignoring the stars because of the likely assimilation and exaggeration. The identification seems certain. The place will be taken as $37^{\circ}.4N$, $41^{\circ}.8E$, with standard deviations of 2° in both latitude and longitude. This is an arbitrary estimate of the amount of assimilation.

This eclipse was also reported from China (360 Aug 28 C).

418 Jul 19 M. Ginzel reproduced and translated a passage from the contemporary historian Philostorgius,

which says approximately: " . . . while Theodosius was
in his minority and the month of July had advanced to the
19th day, the sun was eclipsed so strongly about the eighth
hour that stars could be seen." The passage goes on to
describe a terrible drought which followed the eclipse. It
then returns to the eclipse and says that a "cone-shaped
light" appeared in the sky during the eclipse, which some
took to be a comet.

In spite of its magical trappings, this report is probably reliable. Ginzel says that Philostorgius used eye-
witness accounts by contemporaries for his history after
about 395. The month and day are correctly given. The
minority of Theodosius (as emperor) lasted from 408 to 421
and there was an eclipse on the day given in the year 418.
The hour of the day is reasonable for this eclipse. If the
detail about the comet is accurate, it is the earliest record
I have seen of a comet appearing during an eclipse. Because of the comet, I shall take the eclipse to be total.

Unfortunately, the place is not well known. Ginzel
takes the two possible places of observation to be
Constantinople (41°.0N, 29°.0E) and Borissus, which he
places at 39°.9N, 36°.7E. I shall use these.

There are several other records of this eclipse.
All seem to come from one or the other of two independent
sources. The other source was in the Iberian peninsula
and appears in the report 418 Jul 19 E.

484 Jan 14 M. The Athenian philosopher Marinus
Neapolitanus[†] wrote a biography of the Athenian philosopher Proclus (412-485). Ginzel reproduces a passage from
it, along with a translation into German, which reads approximately: "There were also omens for a year before
his death, for example, an eclipse of the sun which was so
great that it made night out of day. A deep darkness began
and stars could be seen. This happened in the sign of

[†]From the town of Shechem in Samaria; Emperor Vespasian
renamed it Neapolis. It is the present Nablus.

Capricorn in the eastern sky[†]. " Marinus also notes that there was another eclipse predicted, which "will happen at the end of the first year" (after Proclus' death).

The year of the large eclipse is given closely, and the identification seems certain. Until 484 Jan 14, the last eclipse that was probably visible in Athens was that of 472 Aug 20, which Oppolzer's Canon shows as passing across northern Africa. Further, the eclipse of 484 occurred near sunrise at Athens while that of 472 was near noon.

In spite of the magical aspect, the report has high reliability. Marinus was a student of Proclus and succeeded him in the chair of philosophy in Athens. Since he referred to the eclipse of the first year after Proclus' death (this eclipse was probably that of 486 May 19) as "predicted", he must have written the biography in 485 or early 486, and there is thus strong reason to believe that he was an eye witness of the eclipse of 484. Because of the magical aspect, I shall lower the reliability[‡] to 0.5. I shall take the place to be Athens, and shall use 0.01 for the standard deviation of the magnitude.

590 Oct 4 M. Ginzel gives, in parallel columns, a quotation and a translation from the contemporary Byzantine historian Theophylactus Simocatta (V, 16) which reads approximately: "In spite of all dissuasion, Caesar Mauricius marched out of his palace and proceeded 1 1/2 parasangs to the hebdomon[*], as it is called in Byzantium. On that day a very large eclipse of the sun occurred; that was the ninth year of the Caesar Mauricius. Then a windstorm arose suddenly, and an overpowering south wind that almost threw the gravel from the bottom of the sea in its rage. "

[†] Hence the eclipse was probably in the early morning.

[‡] The detail that the sun was in Capricorn helps confirm the date of 484 Jan 14. Marinus could have known this detail from the date; in fact, it may be his way of stating the month.

[*] Apparently the seventh milestone.

Superficially this looks like a magical eclipse, but it probably is not. Other things must sometimes happen at the same time as a solar eclipse. Nothing is made of the eclipse as an omen. Mauricius reigned for another dozen years, and the campaign he started was successful. Since there was a large eclipse during the year stated, I give this report a reliability of unity. The identification cannot be questioned unless the year is wrong by two years. The place is clearly Constantinople. The standard deviation of the magnitude will be taken as 0.06.

Gregory of Tours, from the other end of Europe, recorded this as a partial eclipse, but not with usable accuracy in time or magnitude.

6. ECLIPSES THAT WILL NOT BE CALCULATED

Tables IV.1 through IV.5 contain only eclipses whose circumstances will be calculated for Part II of this study. For several eclipse reports discussed in Chapter IV the circumstances will not be calculated and there is no corresponding tabular entry. Such reports include some well known ones that many readers will expect to find listed. Table IV.6 has been prepared to help the reader who does not find an expected listing in the earlier tables.

TABLE IV. 6

ECLIPSE REPORTS THAT WILL NOT BE USED FURTHER

Date Used to Identify Discussion	Section of Chap. IV	Associated Name; Conclusions
ca -2000	3	Chung K'ang or the drunk astronomers; myth concerning failure of a calendrical cycle and the necessary magic ritual.
-699 Aug 6	2	Unidentifiable.
-556 May 19	5	Larissa; darkness not an eclipse.
+ 29 Nov 24	5	Phlegon; magical eclipse that could not be located if real.
83 Dec 27	5	Plutarch; literary eclipse that could not be located if real.
186 Dec 28	4	Occurred at sunset, making magnitude unknowable.
218 Oct 7	5	Cannot be located.

121

The first column in Table IV.6 gives the date used in connection with the discussion, the second column gives the section of Chapter IV in which the discussion is found, and the last column gives the traditional name associated with the report, if any, and the main conclusion reached in the discussion.

Between Table IV.6 and the other tables in Chapter IV, every report that received a paragraph or more of discussion is listed. Reports that were dismissed in a sentence are not listed. The three reports in Table IV.6 that have no associated name were used by Ginzel [1899].

Reports listed in Table IV.6 and those listed with zero reliability in other tables share the characteristics that they will not affect the inference to be made in Part II. They correspond to different orders of zero reliability, so to speak. The distinction between those in Table IV.6 and those with zero assigned reliability is not highly important, and the reader should not expect that the distinction has been made on the basis of carefully applied principles.

CHAPTER V

MEASUREMENTS OF ECLIPSE MAGNITUDES AND TIMES

1. COMMENTS ON MEASUREMENTS OF TIME AND MAGNITUDE

Measurements of times and magnitudes[†] of eclipses with enough accuracy to be useful come only from technical reports. Such reports do not need the extensive textual criticism that was needed with many reports of large or total solar eclipses. It is necessary to be sure of the meaning of the technical terms used, to have some idea of the accuracy to be expected from the measurements, and to watch for possible errors in transmission of the record. Some transmission errors are clear from the records themselves. Others will be found only by gross incompatibility with computed results in Part II of this study.

The time measurements record the times when particular phases of an eclipse were reached. The accuracy of a time measurement involves both the accuracy of choosing an instant when a phase was reached and the timing error proper involved in assigning a value of the time to the chosen instant.

a. Meanings of the phases and the accuracy of their determination. Phases and times connected with both solar and lunar eclipses are listed in Tables V. 2, V. 4, and V. 5, in a form that is as close as feasible, for a modern English text, to the oldest available form of the record. Almost all the phases are listed as "beginning", "middle", and "end".

[†] The statement about magnitudes applies only to partial eclipses.

123

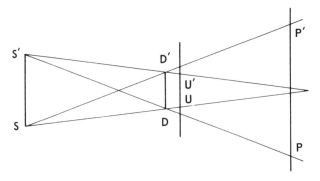

Figure V. 1. Schematic representation of the assumed meaning of the beginning and ending phases of eclipses.

The meanings that will be attached to the beginning and end phases can be seen with the aid of Figure V. 1. In this figure the line SS' represents the diameter of the sun and the line DD' represents the diameter of the body producing the eclipsing shadow. For a solar eclipse DD' is the moon, and the observer can be thought of as moving upward along the line PP'. For a lunar eclipse, DD' is the earth, and the moon can be thought of as moving upward along the line UU'; the observer is on the dark side of DD'.

For a solar eclipse it is plausible that the beginning and end phases mean that the observer is at P and P', respectively, that is, at the points where he enters and leaves the penumbra. At any time that the observer is within the penumbra of the moon, he is unable to see some portion of the sun's disc.

Definition of the beginning and end phases for a lunar eclipse is more difficult. I shall assume that they mean that the leading limb of the moon is at U and that the trailing limb is at U', respectively. That is, between the beginning and end phases as assumed, some portion of the lunar disk is assumed to be within the umbra of the earth. This contrasts with the assumption made for solar eclipses, which was concerned with the penumbra. The

124

assumption that U and U' are the critical points seems plausible for an observer who uses only the unaided eye. As the leading limb of the moon crosses the penumbra, it continues to receive illumination, although in decreasing amount, until it reaches U, and the observer would be able to see the entire circle of the moon's rim until U is reached. However, the existence of the penumbra certainly complicates the task of deciding when U is reached.

I have not found any information on the ability to judge the beginning and end of an eclipse with the unaided eye. The ability probably depends upon the observer, upon circumstances such as cloud cover, and upon the body being eclipsed. In order to gain some information I cut three disks 10 cm in diameter from white paper. One disk was thereafter left intact. The others were cut approximately to the shapes corresponding to eclipse magnitudes of 0. 005 and 0. 01 of the diameter. (Magnitude will be defined later in this section.) I illuminated the disks in front of a dark background and viewed them from a distance of 10 meters, singly and in various combinations. Without knowing the identity of the disks in advance, I was always able to identify the "eclipse" of 0. 01 magnitude and was unable to distinguish the other two even when they were presented to view simultaneously.

This experiment is far from definitive, but elaboration of it does not seem warranted in view of the other uncertainties involved in using the ancient eclipses. For the purposes of this chapter, then, it will be assumed that the "beginning" and "end" phases of an eclipse mean the phases at which 0. 0075 of the diameter is eclipsed. A standard deviation of 0. 0025 will be assigned to the judgment of the phase.

This assumption means, among other things, that the beginning is judged too late, on the average, and that the end is judged too early. If the error is symmetrical, as is assumed, the size of the error would not matter provided the beginning and end phases were observed an equal number of times. However the beginning is represented 32 times in the tables and the end only 20 times. A wrong assumption about the meaning of the observed phases thus matters. The sensitivity of the results to the precise interpretation placed upon the phases will be tested in the analysis by two computations. In one computation all observations will be given equal weight. In the other the collection of beginning phases will be given the same weight as the collection of end phases.

The phase called "middle" could conceivably mean many things, such as the greatest phase, the average of the beginning and end phases, or the occurrence of syzygy with respect either to right ascension or longitude. Ptolemy [ca 152; VI.5] defines the time of the "middle" to be the average of the times of beginning and end. In modern almanacs the middle phase still means an average. The times that are averaged are now defined more precisely, but the distinctions are negligible here. For these reasons it will be assumed that "middle" refers to the average of beginning and end[†].

If the time of the middle phase were measured by first measuring the times of the beginning and end phases and then averaging, the precision of the middle phase would be slightly better than that of either of the other two. If it were actually measured by judging a time of symmetry

---

[†] The record -173 Apr 30 M confirms this assumption. The record 364 Jun 16 M contradicts it. This record gives times for beginning, middle, and end, and the time for the middle is not the average of the other two. No other record gives all three of these phases, so there is no other test of the assumption. Fotheringham [1920] took "middle" for the eclipse of 364 to mean "greatest magnitude", but that assumption did not fit well either. Transmission error in one of the numbers, or an attempt to judge the middle by eye, may be the explanation.

Further, Ptolemy, in Book V.14, contradicts the assumption and equates "middle" with "maximum" in connection with the record -620 Apr 22 BA. It is indeed fortunate, as is mentioned later in the text, that it is not important to know what "middle" meant.

rather than by strict application of the assumed definition, its precision would probably be less. For simplicity, the same accuracy will be assumed for all three phases.

On the average over all geometries of eclipses, all likely definitions of the middle phase will give the same result for the accelerations of the earth and moon. Thus, while the assumed meaning of "middle" is not solidly based, we do not depend as critically as one might think upon the correctness of the assumption.

A few phases besides beginning, middle, and end occur in Table V. 4 for the Islamic data. "At 3 digits" and "at 6 digits" will be obvious after "digit" is defined later in this section. "Greatest" occurs for two partial solar eclipses. We may presume that the astronomer meant what he said, but we have no way to know how he made the measurements; these measurements will be ignored. "Opposition" occurs once, but we do not know what kind of opposition; it is simple and costs little to ignore this observation. The Arabic term that Caussin [see Ebn Iounis, 1008] rendered by "attouchement" occurs twice. In the record 990 Apr 12 Is, "attouchement" seems synonymous with "beginning". In the record 986 Dec 19 Is, the terms are definitely distinguished, but there is reason (see Section V. 4) to think that this record has been garbled. It is probable that "attouchement" means the same as "beginning" but, for safety, the times of "attouchement" will be ignored. Finally, in the record 1004 Jan 24 Is, we have a phase called "beginning" followed by a phase called "sensible to view". Newcomb [1875] used the first phase but not the second; I consider it safer to ignore both because of the uncertain meaning.

b. The accuracy of time measurements. The time measurements fall into three groups listed in Tables V. 2, V. 4, and V. 5. Each group is separated widely in time and place from the others and is expected to have a measurement accuracy that is different from the others. The accuracy of timing for each group will be discussed in the appropriate section of this chapter. Each group is spread

127

over a considerable time, and thus there is a possibility of change in accuracy within a group. This possibility will be ignored.

The detailed calculations to be made in Part II of this work will furnish an a posteriori estimate of the resultant error, for each group, from errors in timing and in estimating phase. The a priori estimates to be made in this chapter should be useful in judging reasonableness and in detecting errors in the transmission of the records.

c. The definition of eclipse magnitude and the accuracy of its measurement. The magnitude of an eclipse is defined in two different ways, which will be called "magnitude of the diameter" and "magnitude of the area".

The magnitude of the diameter means the fraction of the diameter of the bright body that is obscured from view. For example, one of the sources of the eclipse of Hipparchus (see the record -309 Aug 15b M, Section IV.5) says that the sun, as seen at Alexandria, "was eclipsed with the exception of one-fifth of its own diameter..." Since 0.2 of the diameter was not eclipsed, 0.8 of the diameter was eclipsed and the eclipse magnitude was 0.8 of the diameter.

The magnitude of the diameter sounds like an easy quantity to measure, and one can readily imagine simple instruments that ancient astronomers could have made that would allow rather accurate measurement of the magnitude of the diameter. It is something of a surprise to find that Ptolemy [ca 152, VI.7] says about measuring magnitudes: "But as most of those who observe eclipses do not measure the magnitudes of the shadowing by the diameters of the disks but by their entire surfaces, comparing grossly by sight what they see of the disk with what they do not see, we have added to these tables a shorter table . . ." That is, in or before Ptolemy's time, most eclipse magnitudes were given as the fraction of the disk area of the bright body that was obscured. This fraction will be called the "magnitude of the area". The table

128

mentioned in the passage quoted is a conversion table connecting the magnitude of the diameter and the magnitude of the area.

TABLE V. 1

PTOLEMY'S TABLE FOR CONVERTING ECLIPSE
MAGNITUDES FROM DIGITS
OF DIAMETER TO DIGITS OF AREA

Digits of Diameter	Digits of Area of the Sun	Digits of Area of the Moon
1	1/3	1/2
2	1	1 1/6
3	1 3/4	2 1/15
4	2 2/3	3 1/6
5	3 2/3	4 1/3
6	4 2/3	5 1/2
7	5 5/6	6 3/4
8	7	8
9	8 1/3	9 1/2[a]
10	9 2/3	10 1/3
11	10 5/6	11 1/3
12	12	12

[a]Should perhaps be 9 1/3

Ptolemy's table is reproduced, using Arabic numerals but otherwise giving the fractions in the form he used, in Table V. 1. The functional relation between the two measures depends upon the relative sizes of the shadowed and the shadowing bodies and is hence different for the sun and moon. Since the apparent sizes of the sun and moon are not the same for all eclipses, the relation differs slightly for different eclipses. Ptolemy has presumably calculated Table V. 1 for what he believed to be the average apparent sizes; this table is certainly good enough for present purposes.

In Table V. 1 the magnitude, whether of the diameter or the area, is expressed in "digits". In this context a digit means one-twelfth. The digit is used to express all Babylonian magnitudes, all Greek magnitudes with one exception, and all Islamic magnitudes with one exception[†].

[†] With the understanding that the magnitude, when regarded as a fraction, is sometimes reduced to simplest form

129

In the Chinese records, on the other hand, the magnitude, with one exception, is given in units of a fifteenth.

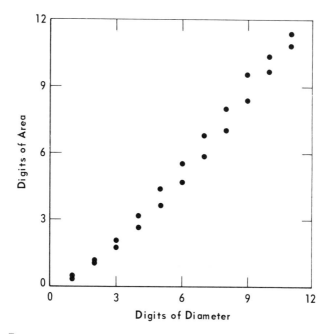

Figure V.2.   Ptolemy's relation between the magnitude of the diameter and the magnitude of the area for solar and lunar eclipses. The point for a lunar eclipse lies above the point for a solar eclipse, for each value of the magnitude of the diameter. The unit called a digit is a twelfth part. The value for a lunar eclipse at 9 digits of diameter has probably been copied wrong at some stage of transmission.

The values listed in Table V.1 are plotted in Figure V.2 but are left unconnected. For a given magnitude of the diameter the magnitude of the area is always greater for a lunar than for a solar eclipse. It can be seen from the figure that $9\frac{1}{2}$ digits of area for a lunar eclipse, when the

instead of being given in digits. For example, 3 digits is sometimes replaced by 1/4. With the exceptions noted, when a fraction is used, its denominator is always a factor of 12 (or of 15 for the Chinese records).

magnitude of the diameter is 9 digits, does not fit smoothly with the other points. It is probable that the value was originally $9\frac{1}{3}$ digits and that this was copied as $9\frac{1}{2}$ at some stage of the transmission.

It is to be expected that the accuracy of measuring the magnitude is different for each group of eclipses, just as the accuracy of measuring time is expected to be. However, it is convenient to discuss the accuracy of measuring the magnitudes collectively here, except for the Chinese measurements, which will be discussed in Section V. 3.

It is reasonable to expect that an experienced observer or set of observers will have a good idea of the accuracy of a measurement and that it will be comparable with the least significant figure given. Thus, in the Mediterranean (Greek) group of records in Table V. 5, the magnitudes are all given to an integral number of digits of the diameter. In the Islamic group of records in Table V. 4, the magnitude is given as an integral number of digits in 9 cases and is given to greater precision than a digit in 15 cases; there is also the exceptional case already mentioned, 854 Feb 16 Is, for which the magnitude is not measured in digits at all. Neglecting the exception, the distribution of significant figures is about what we would expect if the magnitude were always measured to a precision of about $\frac{1}{2}$ digit.

It is reasonable to assume a priori that the Islamic measurements are accurate to the figures given; that is, that the maximum error is about 0. 25 digit and hence that the standard deviation is 0. 14 digit or about 0. 012 in modern decimal notation. It is reasonable to assume a priori that the Greek group is accurate to 0. 5 digit maximum error, or to 0. 29 digit or 0.024 standard deviation.

There is confirmation of this estimate for the Greek group. Fotheringham [1909b] used the magnitudes listed in Table V. 5 to obtain an equation of condition connecting the accelerations, and found that the residuals of the magnitudes ranged in absolute value from 0. 01 to 0. 04; the latter value is just under 0. 5 digit.

131

It is more difficult to make an accuracy estimate for the Babylonian group in Table V.2. First there is the question of whether the measured magnitudes apply to the diameter or the area. The passage from Ptolemy quoted earlier in this section would lead one to think that measurements as old as those in Table V.2 would be of the area. However, Ptolemy explicitly uses the magnitudes of the eclipses of -620, -522, and -501 as magnitudes of the diameter. There is no clue in the record about the other magnitudes, and I have taken them to be of the area because of Ptolemy's statement that most magnitude measurements are of area.

If the magnitude of the area were measured grossly by visual comparison of the uneclipsed and the eclipsed areas, as Ptolemy says, we may reasonably infer that the magnitude of the area was, at least originally, measured without instrumental aid and that it was probably an earlier measure than the magnitude of the diameter. It is thus a plausible hypothesis that the Babylonian magnitudes, which were measured between -720 and -382, were originally of the area and that Ptolemy converted some of them to diameter, by means of Table V.1, before using them in his geometric calculations. If this be so, we expect that the errors in the Babylonian magnitudes are larger than those in the Greek ones for two reasons. First, measuring the magnitude of the area is a gross measurement, according to Ptolemy. Second, the magnitude of the area, already given only to a digit, was converted to the magnitude of the diameter and the new result rounded to an integral digit.

Even if the hypothesis is not correct, it is still likely that there are larger errors in the Babylonian magnitudes than in the Greek ones, because they are from 2 to 5 centuries earlier.

I shall assign an a priori standard deviation of 0.6 digit or 0.05 in decimal notation to the Babylonian magnitudes.

  d. <u>The accuracy of the measured magnitude in the eclipse of Hipparchus</u>. We are now in a position to assign an error estimate to the measurement of the magnitude at Alexandria in the report of the eclipse of Hipparchus (see report -309 Aug 15b M, Section IV.5). It should be recalled that the eclipse in question had a magnitude of the diameter of about 4/5 or 9.6 digits.

  <u>Fotheringham</u> [1909b] used an estimate of the accuracy to justify identifying the eclipse as that of -128, and used some circular reasoning in doing so. He started with the conclusion already quoted that the largest error in the magnitudes given in Table V.5 is slightly less than $\frac{1}{2}$ digit. He also stated[†] that "the eclipse of Hipparchus cannot be identified with any except that of -128 Nov 20 unless a large error - in the case of Mr. Cowell's formulae, three-quarters of a digit - is assumed in the observed magnitude at Alexandria." Fotheringham then assumed that the accuracy of $\frac{1}{2}$ digit should apply to the eclipse of Hipparchus and hence concluded that the eclipse was that of -128.

  This is circular reasoning because the assumed accuracy is valid only within the time span of Table V.5, from -173 to +136. The eclipse of -128 is the only possibility within this span. The estimate does not apply to earlier possible identifications. The estimate, however, which tacitly assumes the date of -128, is used to eliminate possibilities to which it does not apply.

  For example, assume that the eclipse was that of -309 Aug 15. For this date, it is reasonable and in fact probable that the measurement error is greater than the

---

[†]Fotheringham quotes this statement from a paper by Cowell, <u>Monthly Not. Roy. Astro. Soc.</u> <u>69</u>, 617-618, 1909, that I do not have. Since Fotheringham adopted the statement, it does not matter for the present purposes what Cowell actually said, and I have not verified the reference.

$\frac{1}{2}$ digit expected in -128. An error of $\frac{3}{4}$ digit for the eclipse of -309 is therefore just as consistent with the record as an error of less than $\frac{1}{2}$ digit for the eclipse of -128[†].

In studying the meaning of the magnitude measurement at Alexandria, no one seems to have commented upon the odd way in which the magnitude is stated. This is the only case between at least the years -720 and +136 in which the magnitude is reported in any way other than as an integral number of digits or the equivalent. This fact raises a suspicion that the magnitude was originally reported in digits and was changed to the present form by a later editor.

From Table V. 1 or Figure V. 2 we see that 9 digits of area, for a solar eclipse, equals $9\frac{1}{2}$ digits of diameter. Thus it is a plausible assumption that the magnitude in the original record was 9 digits of the area, and that a later editor first changed this to $9\frac{1}{2}$ digits of the diameter[‡] and then approximated this by "about four fifths".

In using the magnitude at Alexandria, I shall assume that this is what has happened and that the correct magnitude is $9\frac{1}{2}$ digits of the diameter, or 0.79 in decimal notation. This value will be given a standard deviation

---

[†] Further, Fotheringham [1920] contradicted one of the basic premises of Fotheringham [1909b], apparently without noticing it. In 1920, Fotheringham calculated for the eclipse of -309 that the magnitude at Alexandria could have been 9.2 digits for totality somewhere on the Hellespont. This is a discrepancy of only 0.4 digit, which is well within the specified accuracy of $\frac{1}{2}$ digit, contrary to Fotheringham's 1909 premise. Since the record does not require totality on the Hellespont, the magnitude at Alexandria could be even closer to 9.6 digits.

[‡] Note that the magnitude of the diameter is what Hipparchus needed in his use of this eclipse, so he must have stated the magnitude of the diameter at some point. He, or some later writer, may have thought it was pointless to preserve any other statement of the magnitude.

134

appropriate to the measurement of magnitude of the area, namely 0.05 in decimal notation.

At least initially in the computations to be carried out in Part II of this work, the magnitude at Alexandria will be used only to see if the number of possible identifications of the eclipse can be reduced below those given in Section IV.5. Unless the eclipse can be identified it does not seem worthwhile to use the probably inaccurate measurement of magnitude. Prospects for identification do not look bright at this point in the investigation.

It is doubtful that the identification will be affected by use of the magnitude as 9 digits of the area rather than as 4/5 of the diameter. The interpretation of the magnitude as originally a measurement of the area is advanced mainly to explain a peculiar feature of the eclipse report.

2. BABYLONIAN MEASUREMENTS OF ECLIPSE MAGNITUDES AND TIMES

Ptolemy [ca 152] has transmitted measurements, presumably made at Babylon, concerning ten lunar eclipses. The validity of these records has been attacked, and it is desirable to consider this point first.

It is interesting to note that the validity or quality of almost all the data transmitted by Ptolemy has been questioned by some one. No one defends the accuracy of his equinoxes and his star tables. Fotheringham [1915a] stated that Hipparchus' equinox measurements, which Ptolemy transmitted, were badly made and used an argument based on this claim to explain Ptolemy's bad value of the precession; later [Fotheringham, 1918] he upheld the accuracy of Hipparchus' equinoxes and had a different explanation of the wrong precession. Newcomb [1875] rejected the validity of the occultations (see Section VI.1) while Fotheringham [1915a and 1915b] upheld it. They interchanged positions with regard to the Babylonian eclipses, which Newcomb used and Fotheringham did not.

TABLE V. 2

BABYLONIAN MEASUREMENTS OF TIMES AND MAGNITUDES
OF LUNAR ECLIPSES

Identification	Chapter[a]	Magnitude, Digits and Side Eclipsed		Phase	Time at Babylon
		Area	Diam.		
-720 Mar 19 BA	IV. 5		Total	Beg.	More than 1^h after moonrise
-719 Mar 8 BA	IV. 5	3S		Mid.	Exactly midnight
-719 Sep 1 BA	IV. 5	> 6N		Beg.	After rising
-620 Apr 22 BA	V. 14		3S	Beg.	End of 11th hour
-522 Jul 16 BA	V. 14		6N	Mid.	1^h before midnight[c]
-501 Nov 19 BA	IV. 8		3S	Mid.	2/5 hour before midnight
-490 Apr 25 BA	IV. 8	2S		Mid.	Middle of sixth hour
-382 Dec 23 BA[b]	IV. 10	1 ? NE?		Mid. ?	1/2 hour before sunrise
-381 Jun 18 BA[b]	IV. 10	? NE?		Mid. ?	1 hour of the night
-381 Dec 12 BA[b]	IV. 10	Total		Beg.	After 4^h of the night

[a] Book and chapter number in Ptolemy [ca 152].

[b] Will not be used in final inference. See text.

[c] Fotheringham [1915] gives 3 1/3 hours after sunset. See text.

A cuneiform text [Kugler, 1909, p. 71] has been found that refers to the lunar eclipse of -522 Jul 16 (see also Table V. 2). It gives the time of the eclipse as 1 2/3 double-hours after sunset. The calculated time of sunset at Babylon for the date in question is $19^h 02^m$ local apparent time. Fotheringham [1915b] added $3^h 20^m$ (= 3 1/3 hours = 1 2/3 double-hours) to obtain $22^h 22^m$ for the time in the cuneiform record. The record of the same eclipse transmitted by Ptolemy says 1^h before midnight[†]; it is almost but not absolutely certain that this is the time of the middle. On the assumption that Ptolemy's record and the cuneiform record refer to the same phase, Fotheringham found a discrepancy of 38^m between them. On the assumption that the cuneiform record refers to the beginning and Ptolemy's to the middle, he still found a discrepancy of 38^m because the interval between the phases should be 76^m for this eclipse, according to Oppolzer's Canon.

[†] In the quotation of the record, Ptolemy did not say whether the hour was equal or unequal. In computations dealing with the eclipse, he took it as an equal hour, and I see no reason not to adopt this interpretation. I also accept his interpretation of the phase as the middle.

Fotheringham thereupon rejected all Babylonian eclipse records on the basis that they had been "reduced" before reaching the form that Ptolemy gave. In my opinion, an error in one record does not provide sufficient grounds for rejecting all records from an entire provenance. Further, if Fotheringham chose to reject "reduced" data, he should have rejected the Greek as well as the Babylonian records. There is little question that both the Greek and the Babylonian data have been reduced. Ptolemy's statements show this clearly. For example, in the record called -200 Sep 22 M in Table V.5, Ptolemy states that an eclipse of the moon began half an hour before the moon rose. This statement cannot be "raw data". The meaningful question for this study is not whether the data have been reduced. It is not even whether the reduction has been done accurately, within reasonable limits. It is whether the reduction has been done in a way that would be likely to bias our results.

In the case that Fotheringham discussed, the reduction error may not be as bad as he thought. Let us assume that the time in the cuneiform record was in unequal hours. During the night of -522 Jul 16 at Babylon, 1 un. hr. = $49^m.6$ approximately. Hence the time in the cuneiform record would be $19^h 02^m + (3\,1/3) \times (49^m.6) = 21^h 47^m$ local apparent time rather than $22^h 22^m$. One eq. hr. before midnight is of course 23^h. The interval would be 73^m rather than 38^m and the discrepancy would be 3^m rather than 38^m. An error of this size would be negligible[†].

[†] An explanation in the literature needs comment. It has three main steps: (a) a cuneiform inscription sometimes called the "ivory prism" carries an inaccurate table for converting equal to unequal time units; (b) a Babylonian astronomer used the table to convert 1 2/3 equal double-hours into unequal hours; and (c) Ptolemy ultimately obtained the converted record rather than the original. This explanation increases the discrepancy. It rests upon a particular interpretation of the table on the ivory prism. The columns in the table are not labelled, and it is not certain that they deal with equal and unequal time. For example, Neugebauer [1947] thinks that the table deals with

It does not need to excite comment that Ptolemy gave the time of the middle phase in equal hours if the cuneiform record gave the time of beginning. Ptolemy used the middle, presumably as an approximation to opposition, in equal hours in his calculations with all the eclipse records; sometimes he gave the details of how he found the middle time and sometimes not. It is marginal whether he would have tried to preserve an accuracy of 3^m in his calculations and statements. Thus it seems plausible to me that the calculation just outlined is the one that Ptolemy used. Whether it is correct that "1 2/3 double-hours" was in unequal time is another matter[†]; Ptolemy could have misread the record.

The possibility that the cuneiform time was in unequal time is denied implicitly by Fotheringham [1915b] and explicitly by Kugler [1909, p. 63], Neugebauer [1955, v. 1, p. 39], and van der Waerden [1956, p. 88]. I do not understand the basis for so much assurance. Back of it are assumptions about uniformity of human conduct and uniformity of word usage at all times in all contexts that go beyond my experiences of human consistency. A corollary of the assurance is that the cuneiform text in question can be read with no ambiguity in the technical terms involved. We can test this corollary by comparing independent translations of the (transcription of the) cuneiform record. Kugler and Sachs[‡], respectively, give the

---

the weight of water needed in a water clock at various seasons. Finally, the errors in the table are unreasonably large if it deals with equal and unequal hours.

[†] It should be noted that a consistent error of misreading equal hours as unequal hours produces a mean error that approaches zero as the sample size increases. Hence the possibility of an error in the kind of time does not justify rejecting the records.

[‡] Kugler's translation appears in [Kugler, 1909, p. 71]. Sachs' translation is contained in a private communication from Prof. A. J. Sachs to Britton and appears in [Britton, 1967, p. 82]. Kugler's translation is into German. It will appear that uncertainties caused by translating this translation are negligible.

following translations, in which the year number is the year of the reign of Cambyses II and Duzu is the fourth month of the Babylonian calendar:

> Year VII, Duzu, night 14, 1 2/3 double-hours after beginning of night, a lunar eclipse; entire course visible; it extended over the northern half of the disk.

> Year VII, month IV, night of the 14th,1 2/3 double-hours in the night a "total" lunar eclipse took place [with only] a little remaining [uneclipsed]. The north wind blew.

When there is this much divergence between excellent scholars, I hesitate to be sure about the precise meaning of Babylonian terms.

Even if the authorities should be able to establish that the Babylonian inscriber intended "double-hour" to be a unit of equal time, they would be unable to show that Ptolemy read it that way.

The difference between assigning the middle of the eclipse to 1 eq. hr. before midnight and assigning the beginning to 1 2/3 equal double-hours after sunset is only 38^m. The difference between ephemeris time and solar time at the epoch in question is probably about four or five hours. An uncertainty of 38^m out of several hours, for one record only, is not important for the purposes of this study. I have spent this much time on the record -522 Jul 16 BA only because the controversy about it bears on the question of whether to accept Ptolemy's versions of the Babylonian records. Since Ptolemy (or some earlier Hellenistic astronomer) was closer in time and place to Babylon than we are, I am as willing to accept his translations (but not necessarily his interpretations) as those of anyone else, in spite of problems about this record.

The identifications of the Babylonian eclipses and the chapter of Ptolemy in which the record is found are listed in the first two columns of Table V. 2. Some of the records are straightforward and need no comment beyond that furnished by the table; some need explicit comment.

-522 Jul 16 BA. This record has already been discussed in part. The text reads ". . at one hour before midnight . . at Babylon one saw the moon eclipsed by half its diameter in the northern part. " As it stands, this is a record of the time when the magnitude reached a particular value. It is not explicitly stated either that the time is the middle or that the phase is greatest. Not until an Islamic measurement on 1004 Jan 24 do we find another record when the times of specific magnitudes were measured. Therefore I assume that the time is that of the middle and that the magnitude is the maximum.

-501 Nov 19 BA. The text reads: "The eclipse . . . occurred at 6 1/3 eq. hr. of this night. The moon was eclipsed by 1/4 of its diameter on the southern side, and the middle was at 2/5 of an hour before midnight for Babylon. " The manner of reporting the time of this eclipse is quite different from the manner used for eclipses surrounding this one in time, so the times have probably been edited. Also the text is not clear about what phase occurred at 6 1/3 eq. hr.

6 1/3 eq. hr. of the night for this date at Babylon is about $23^h 18^m$ local mean time. Midnight apparent time is about $23^h 45^m$ mean time and 2/5 of an unequal hour is about 27^m in equal time, so that 2/5 of an hour before midnight is also $23^h 18^m$ local mean time. The exact agreement is surely fortuitous, but it is clear that only the time of the middle phase is reported. A late editor probably calculated the time measured in equal hours and inserted it into the record.

-382 Dec 23 BA, -381 Jun 18 BA, and -381 Dec 12 BA. These records share several peculiarities. They refer to consecutive eclipses. They are the only records that Ptolemy transmitted from the -4th century, and they are the only Babylonian records in Chapter IV. 10. They are the only Babylonian records whose dates are given in

140

terms of the Athenian calendar. Records from this chapter, whether Babylonian or Greek, are the only ones in which a measured magnitude is not given for partial eclipses (see the discussion of the record -309 Aug 15b M (Hipparchus)). They all use, and are the only ones that use, the phrase "eclipsed from the side of the summer rising". Finally, Ptolemy describes them as "three eclipses out of the number of those brought from Babylon as having been observed there. "

Ginzel [1899] and others doubt that the eclipses were actually observed at Babylon. It has been suggested that they were observed in a Greek city and the records taken to Babylon where the times were converted to Babylon time. Britton [1967, p. 88] suggests that they represent attempts to predict eclipses. If either suggestion is correct, the records could introduce bias into the results, and we must proceed carefully.

van der Waerden [1958] feels strongly that the three eclipses were observed in Babylon and that the first two should receive more than average weight, but I cannot share this feeling. For example, an argument on which he seems to place much weight is that the eclipse of -381 Jun 18 would have had to begin before sunset in Athens or any other likely city of the Greek culture, and hence that the record could not have come from any likely Greek city. However, we cannot conclude that an eclipse time would not be recorded simply because it was not observed. The Greek record -200 Sep 22 M explicitly states that a lunar eclipse began half an hour before moonrise.

The record -382 Dec 23 BA seems internally inconsistent to me if we assume that the observations back of it were made in Babylon. Ptolemy quotes the record as saying, after giving the date, that "a small part of the disk was eclipsed from the side of summer rising[†] when half an hour was left of the night, and the moon set still eclipsed. " It seems to me that this explicitly denies that the observed phase was the beginning; however, Ptolemy assumed that

---

[†] Northeast?

the eclipse began at the time stated. He later said, in connection with finding the middle time, that "since only a small part entered the shadow, the entire duration of the eclipse must have been at most $1\frac{1}{2}$ hours." If the eclipse began at the time stated, it would have been impossible to know by observation either the magnitude or the duration, because of the interposition of moonset and sunrise.

It seems to me either that some or all of the circumstances of the eclipse were calculated, as Britton suggests, or that Ptolemy had a record of the eclipse from somewhere other than Babylon[†], or that the observed phase was the middle. In any case, there are questionable circumstances surrounding the records. It is safest not to use them, since the risk in using them is the introduction of bias, not of mere random error. The eclipses will be calculated, primarily in order to reach a judgment about whether they were observed at Babylon.

The accuracy of the magnitude measurements was estimated in Section V.1. It is also useful to have an a priori estimate of the timing accuracy.

We have no direct knowledge of the way in which the times in Table V.2 were measured. The water clock was almost surely a well known item by the times of the first eclipses in Table V.2. Neugebauer [1947] discussed a cuneiform text that is probably roughly contemporaneous with the first entries in the table. It gives some instructions concerning water clocks; Neugebauer speculated from the instructions that the water reservoir had a constant cross section, so that the rate of flow changed with the hour.

It is a reasonable speculation that a water clock was used to measure the times. If so, we do not know how

[†] Leaving open the possibility that he had a Babylonian record also. Since Hipparchus used these eclipses [Ptolemy, ca 152, IV.10], it may be that Ptolemy had both Hipparchus' statements of the records and additional independent records.

accurate it was. What is worse, we do not know the doctrine of use. There are two plausible doctrines. The clock
for the night could have been started each sunset and the
times read directly from it. This doctrine would make the
error roughly proportional to the interval since sunset. Alternately, the clock could have been started at sunset and
read at dawn and the observations adjusted according to the
dawn reading.

To us it is clear that the latter doctrine is more
accurate, but we do not know whether it was followed or
not. It will be assumed arbitrarily that it was, and it will
be explicitly assumed that the standard deviation of a time
measurement is 10 percent of the interval from sunset or
sunrise, whichever is nearer.

It will be assumed that rounding error in reporting
the time is an additional independent error. Inspection of
the times in Table V. 2 suggests that the Babylonians
rounded to the nearest half hour. The standard deviation
corresponding to this rounding is about 9^m.

3. CHINESE MEASUREMENTS OF ECLIPSE MAGNITUDES
    AND TIMES

Gaubil [1732a, pp. 66-67] gave tables on opposing
pages, one called "calculations of eleven eclipses taken
from the astronomy of the Souy" and the other called "Observations of the same eclipses". The calculations were
made for Siganfou. This is (see Section II. 4) the modern
Sian at 34°. 2N, 109°. 0E. The place of the observations is
not explicitly stated. However, it is stated that the table
of observations is relative to the table of calculations and
that it contains the observations of the calculated eclipses.
Hence the place of observation is presumably Siganfou.
"Souy" presumably refers to the Sui dynasty, which reu-
nited the empire about 590 after some decades of division.

Gaubil's tables give the observed times of several
of the eclipses using Gallicizations of Chinese terms. I
have no confidence in my ability to interpret Gaubil's

terms for time correctly and hence I shall not try to use the times, although they can probably be read by an expert in Chinese astronomy. The observed magnitudes of the six eclipses for which magnitudes were recorded are listed in

TABLE V. 3

CHINESE MEASUREMENTS OF ECLIPSE MAGNITUDES

Identification	Place	Body Eclipsed	Magnitude (fifteenths)
585 Jan 21 C	Sian	M	10
585 Aug 1 C	Sian	S	6
586 Dec 16 C	Sian	S	10
590 Oct 18 C	Sian	M	12
592 Aug 28 C	Sian	M	10
593 Aug 17 C	Sian	M	7 1/2[a]
1221 May 23b C	Samarkand	S	9[b]

[a]Since this is the only eclipse for which the magnitude is given to 1/2 of a fifteenth, and the only one for which the observed magnitude agrees with the magnitude calculated in the Chinese tables, it is probable that this is the calculated magnitude put down by accident.

[b]Actually recorded as 6/10.

Table V. 3. A footnote in the table points out that the magnitude given for the eclipse of 593 Aug 17 is probably the calculated magnitude put down as the observed value by accident. The magnitude recorded for this eclipse will be ignored.

The last line in Table V. 3 gives the magnitude recorded at Samarkand in connection with the record of near-totality for the eclipse of 1221 May 23 (see record 1221 May 23a C). The record [Wylie, 1897] gives the magnitude as six-tenths, which is changed to 9/15 in the table for convenience. This is the only ancient Chinese record in which the magnitude is not given by fifteenths. Since this record is more than six centuries later than the other records, the difference may not be important.

I observed no clue that indicated whether the magnitudes are of the area or of the diameter. The closest thing to a clue occurs with the record 1221 May 23b C. For this eclipse, Wylie's translation says that "six-tenths of the sun was eclipsed" at Samarkand (39°. 7N, 66°. 95E) and that seven-tenths was eclipsed at the unknown place. This sounds more like a measure of area than of diameter. Hence I shall assume that the magnitudes in Table V. 3 are of the area, but the assumption is exceedingly weak.

It will be assumed that the error of observing the magnitude is uniformly distributed with a maximum error of 1/15; this assumption allows somewhat for uncertainty about whether the measurement is of the diameter or the area. This assumption is taken to be equivalent to assigning a standard deviation of about 0. 04 to the magnitude.

4. ISLAMIC MEASUREMENTS OF ECLIPSE MAGNITUDES AND TIMES

Ebn Iounis [ca 1008] has preserved data about many eclipses ranging from 829 Nov 30 to his own time. These data are summarized in Table V. 4. The table gives the data in a form as close to the original as is feasible for a modern English text. Newcomb [1875] also used the eclipse times from Ebn Iounis, but his interpretations occasionally differed from those that will be used here, and he looked only for an acceleration of the moon. The possibility of a nonzero value of $\dot{\omega}_e$ had not yet been seriously considered by astronomers. Newcomb did not use the magnitudes.

Many times in Table V. 4 are given in terms of the altitude or elevation angle of the sun or of a prominent star. Apparently it was standard Islamic practice during this period to fix the time of an observation by means of a measured elevation. Some of the records give both the measured elevations and the times deduced therefrom. When this was done, some of the times agree to a few minutes with the time deduced by Newcomb from the elevations. In other cases, the discrepancy is so gross that a copying error or similar blunder must have occurred. In the absence of gross error attributable to copying, we can put high confidence in the Islamic reductions from elevation angle to time. We can also plausibly assume that all times

TABLE V. 4

ISLAMIC MEASUREMENTS OF ECLIPSE TIMES AND MAGNITUDES

Identification		Place	Magnitude, Digits		Phase	Time
			Area	Diam.		
A. SOLAR ECLIPSES						
829 Nov 30	Is	Bagdad	--	--	Beg.	Ht. of sun 7°
					End	Ht. of sun 24°
866 Jun 16	Is	Bagdad	--	7 1/2	Mid.	7^h26^m un. hr.
					End	8^h30^m un. hr.
891 Aug 8	Is	ar-Raqqah	8^+	--	Mid.	1 un. hr. after noon
901 Jan 23a	Is	Antioch	6^+	--	Mid.	3 2/3 eq. hr. before noon
901 Jan 23b	Is	ar-Raqqah	8^-	--	Mid.	3 1/2 eq. hr. before noon
923 Nov 11	Is	Bagdad	--	9	Mid.	Ht of sun 8° east
					End	2^h12^m un. hr. ; ht. of sun 20°
928 Aug 18	Is	Bagdad	3	--	End	Ht. of sun 11 8/9°
977 Dec 13	Is	Cairo	--	8	Beg.	Ht. of sun 15 1/2°
					End	Ht. of sun 33° 20'
978 Jun 8	Is	Cairo	--	5 1/2	Beg.	Ht. of sun 56°
					End	Ht. of sun 26°
979 May 28	Is	Cairo	--	$5 1/2^a$	Sens.[b]	Ht. of sun 6° 30' west
993 Aug 20	Is	Cairo	8	--	Beg.	Ht. of sun 27° east
					Greatest	Ht. of sun 45° east
					End	Ht. of sun 60° east
1004 Jan 24	Is	Cairo	--	11	Beg.	Ht. of sun 18° 30' west
					Sens.[b]	Ht. of sun 16° 30' west
					At 3 dig.[c]	Ht. of sun 15°
					At 6 dig.[c]	Ht. of sun 10°
					Greatest	Ht. of sun 5°
B. LUNAR ECLIPSES						
854 Feb 16	Is	Bagdad	d	--	Beg.	10^h3^m after noon
854 Aug 12	Is	Bagdad	--	--	Beg.	Ht. Aldebaran 45° 30' east
856 Jun 22	Is	Bagdad	--	8 1/2	Beg.	Ht. Aldebaran 9° 30' east
866 Nov 26	Is[e]	Bagdad	--	1 1/2	Beg.	8^h55^m, un. hr.
					Opp.	9^h31^m, un. hr.
					End.	$10^h7^m30^s$, un. hr.
883 Jul 23	Is	ar-Raqqah	--	10^+	Mid.	8^+ eq. hr. after noon
901 Aug 3a	Is	Antioch	--	$>11 1/2^f$	Mid.	15 1/3 eq. hr. after noon
901 Aug 3b	Is	ar-Raqqah	--	$>11 1/2^f$	Mid.	15 7/12 eq. hr. after noon
923 Jun 1	Is	Bagdad	--	9^+	Mid.	1^h40^m eq. hr. of the night
					End	3^h eq. hr. of the night, ht. α Cygni 29° 30' east
925 Apr 11	Is	Bagdad	--	Total	Beg.	Ht. Arcturus 11° east, 55^m un. hr. of night
					End	Ht. Vega 24°, 4^h36^m un. hr. of night

[a] The text does not make clear whether this was the maximum or the amount at sunset.

[b] "Sensible to view". The meaning is uncertain.

[c] Of the diameter.

[d] Recorded as less than 0.9 of the disk, not 0.9 digit. Assumed to be 10 digits of diameter.

[e] This record has doubtful validity. See Section V. 4.

[f] "A very little less than its diameter" was eclipsed.

146

TABLE V. 4 (continued)

ISLAMIC MEASUREMENTS OF ECLIPSE TIMES AND MAGNITUDES

Identification		Place	Magnitude, Digits		Phase	Time
			Area	Diam.		
B. LUNAR ECLIPSES						
927 Sep 14	Is	Bagdad	--	3 1/2	Beg.	10^h un. hr. after sunset, ht. Sirius 31° east
929 Jan 27	Is	Bagdad	--	Total	Beg.	5^h un. hr. after sunset, ht. Arcturus 18° east
933 Nov 5	Is	Bagdad	--	Total	Beg.	9^h56^m un. hr. after sunset, ht. Arcturus 15° east
979 May 14	Is	Cairo	--	8 1/2	End	1^h12^m eq. hr. of the night
979 Nov 6	Is	Cairo	10	--	Beg.	Ht. moon 64° 30' east
					End	Ht. moon 65° west
980 May 3	Is	Cairo	--	Total	Beg.	Ht. moon 47° 40'
					End	36^m eq. hr. before sunrise
981 Apr 22	Is	Cairo	--	3	Beg.	Ht. moon 21°
					End	1/4 hr. before sunrise
981 Oct 16	Is	Cairo	--	5	Beg. [g]	Ht. moon 24°
983 Mar 1	Is	Cairo	--	Total	Sens. [h]	Ht. moon 66°
					End	Ht. moon 35° 50'
986 Dec 19	Is	Cairo	--	10	Beg.	Ht. moon 24° west
					Att. [i]	Ht. moon 50° 30'
990 Apr 12	Is	Cairo	--	7 1/2	Beg. and Att. [i]	Ht. moon 38°
					End	At rising of 1st deg of Aquarius
1001 Sep 5	Is	Cairo	--	--	End	2^h un. hr. after sunset

[g]Translation cited has "l'attouchement par dehors". [h]"Sensible to view" [i]"Attouchement"

in Table V. 4 were the result of reduction from solar or stellar altitudes.

In order for the altitude measurements to be meaningful, the Islamic astronomers had to measure the altitudes quickly. It is plausible that the accuracy of an altitude measurement made quickly and without a telescope was about 1°. 0. Almost all the altitudes in Table V. 4 are given to a precision of 1°. The corresponding time accuracy would be 4^m if the measurements were of hour angle rather than altitude. Since altitude was actually measured, this value will be doubled and it will be assumed that the standard deviation of an Islamic time measurement was 8^m. This estimate includes the rounding error involved in recording a measurement.

A possible error in the transmission of the data must always be kept in mind. Such an error is in addition to the timing error proper that was just considered. Some transmission errors are obvious from the records themselves; others will probably be found as the result of the detailed calculations to be made in Part II.

The standard deviation of the magnitude measurements has already been estimated (Section V. 1). In the calculations I shall add 0. 1 digit to the magnitudes that have ">" or "+" associated with them in Table V. 4.

For the few reports that do not specify the kind of hour, I shall assume that unequal hours were used.

A few reports need particular discussion.

854 Feb 16 Is. This report contains the only Islamic measurement of magnitude that is not given in terms of the digit. The record says that the "uneclipsed part of the disk exceeded one-tenth". That is, the magnitude was less than 0. 9 "of the disk", or less than 10. 8 digits. From Table V. 1 or from Figure V. 2 we see that $10\frac{1}{3}$ digits of the area for a lunar eclipse is the same as 10 digits of the diameter. For this reason, it will be assumed that the record originally was 10 digits of the diameter.

Newcomb [1875] did not use the time of this eclipse, but I did not notice his reason.

856 Jun 22 Is. The record does not indicate whether the magnitude refers to the area or to the diameter. The observer of this eclipse and the observer of the eclipse of 854 Feb 16 were the same. It was tentatively concluded that the magnitude of the latter eclipse was originally referred to the diameter before editing. For this reason I will assume that this measurement is of the diameter. Luckily, for a lunar eclipse near 8 digits, it makes almost no difference whether the magnitude refers to the area or the diameter (Figure V. 2).

866 Nov 26 Is. The preceding report for 866 Jun 16 gave details of calculations relating to the eclipse first, followed by details of the observations and notes about the discrepancies. This report says that the "opposition should have been" at $9^h 31^m$ and so on. The part of the text giving the numbers in Table V. 4 is followed by remarks that make no sense to me and that give no additional numbers. The presumption is that the numbers listed in Table V. 4 are the calculated ones and that the observed values have been omitted in the available text by some accident.

Newcomb did not use either eclipse of 866, apparently because of suspicion about which values were observed and which were calculated.

883 Jul 23 Is through 901 Aug 3b Is. These reports are from Al-Battani [ca 925] rather than from Ebn Iounis. The magnitudes of the lunar eclipses are specifically stated to be of the diameter. It should be noted that Al-Battani gives reports of the eclipses of 901 from both ar-Raqqah and Antioch (36°. 2N, 36°. 0E). Almost all the diameter was eclipsed in the lunar eclipse of 901 Aug 3; I have arbitrarily interpreted this as a magnitude greater than $11\frac{1}{2}$ digits. The three magnitudes of solar eclipses were described as "according to sight" (secundum visum) rather than as of the diameter. Because of this distinction in phraseology, which is used even with eclipses reported in the same year from the same observatory, I assume that the solar eclipse magnitudes refer to the area. These reports show that identity of observer does not mean identity of usage. Where meaning can be inferred only from the observer's identity and his usage in other situations, my intention is to make the inference weak.

Newcomb [1875] was not acquainted with these records from Al-Battani.

923 Jun 1 Is. A time was given and also the height of "the star near the tail of the Swan". The translator identified this as α Cygni.

149

925 Apr 11 Is. A star elevation and its conversion into time were both given here. There is good agreement for the end phase. According to Newcomb the elevation given for Arcturus yields a time almost an hour before sunset and quite different from the hour given. The hour given is reasonable, so the presumption is that the star identity was recorded wrong.

927 Sep 14 Is. The magnitude is described as greater than $\frac{1}{4}$ of the diameter and less than $\frac{1}{3}$; this is entered in the table as $3\frac{1}{2}$ digits. Similar phraseology was used in almost all cases where the tabular magnitude in Table V.4 is reported to half a digit.

928 Aug 18 Is. The times in the original record were given to a precision of seconds or even of "thirds"; hence they were probably calculations from tables made to compare with the observed circumstances. The height of the sun at the end of the eclipse seems to record a genuine observation. The record says that the eclipse was followed by watching it in a reflecting pool.

979 May 28 Is. The time given was not stated to be the beginning of the eclipse but rather the time when it became "sensible to view". This time will be ignored. It is not clear whether the magnitude given was the maximum of the eclipse or the magnitude at sunset; it will be necessary to test both possibilities in Part II.

980 May 3 Is. The record does not say whether the moon was to the east or to the west at the beginning, but it should not be hard to distinguish these two options if the measurement is otherwise valid. Newcomb [1875] says that the altitude given was not attained at any time during this night and hence that there is a transmission error in the record.

981 Oct 15 Is. Since the phase given was "attouchement", the time in this record will be ignored.

983 Mar 1 Is. The first phase is called "sensible to view", so the time of this phase will be ignored. According

to Newcomb, the altitude of the moon given for this phase is physically impossible, so that the time is not usable anyway.

986 Dec 19 Is. The heights of the moon at the beginning and at "attouchement" differ by 26°. This seems impossible, whatever attouchement may have meant. Further, attouchement would have had to occur before the beginning. It seems likely that parts of different records have been copied by accident. Since the moon set still eclipsed, the altitude listed for the beginning may be correct, and Newcomb used it. I consider it safer to ignore this record altogether.

990 Apr 12 Is. Because "attouchement" is used, the first time given will be ignored. Newcomb [1875] also ignored the time of the end phase, but I did not note his reason.

1002 Mar 1 Is. The translation of Ebn Iounis cited in the references lists a total eclipse of the moon for this date. A translator's note also says that the record is partly garbled and that the translator has restored one of the numbers by calculation. I have omitted this eclipse from Table V. 4.

1004 Jan 24 Is. This record gives times for both the beginning and being sensible to view. These times will be ignored, as will the time of greatest phase of this record and elsewhere. The record does not say whether the magnitude is of the diameter or the area. Luckily, the difference is ignorable for a solar eclipse this large. It will be necessary in Part II to test for the point of observation lying on both sides of the zone of totality, and to hope that one possibility can be excluded on a "reasonableness" basis.

151

5. MEDITERRANEAN (GREEK) MEASUREMENTS OF ECLIPSE MAGNITUDES AND TIMES

Table V. 5 is to be used in connection with this section. The records fall into three main groups.

TABLE V. 5

MEDITERRANEAN (GREEK) MEASUREMENTS OF ECLIPSE TIMES AND MAGNITUDES

Identification'		Place	Body Eclipsed	Magnitude[a]. Side Eclipsed	Phase	Time
- 200 Sep 22	M	Alex. ?	M	--	End	2 1/2 hr. of the night
- 199 Mar 19	M	Alex.	M	Total	Beg.	5 1/3 hr. of the night
- 199 Sep 12	M	Alex.	M	Total	Beg.	6 2/3 hr. of the night
					Mid.	8 1/3 hr. of the night
- 173 Apr 30	M	Alex.	M	7N	Beg.	7 hr of the night
					End	10 hr. of the night
- 140 Jan 27	M	Rhodes	M	3S	Beg.	4 hr. of the night
+ 125 Apr 5	M	Alex.	M	2S	Mid.	3 3/5 eq. hr. before midnight
133 May 6	M	Alex.	M	Total	Mid.	3/4 eq. hr. before midnight
134 Oct 20	M	Alex.	M	10N	Mid.	1 eq. hr. before midnight
136 Mar 6	M	Alex.	M	6N	Mid.	4 eq. hr. after midnight
364 Jun 16	M	Alex.	S	--	Beg.	2 5/6 eq. hr. after noon
					Mid.	3 4/5 eq. hr. after noon
					End	4 1/2 eq. hr. after noon

[a]In digits of the diameter.

The first group contains five lunar eclipses dating from -200 Sep 22 through -140 Jan 27; these are in the time of Hipparchus or somewhat sooner. In the last three of these, the times are explicitly stated to be in unequal hours. It will be assumed that the times for the first two are also in unequal hours. These records have been preserved to us only by Ptolemy [ca 152].

The second group contains four lunar eclipses dating from +125 Apr 5 through 136 Mar 5; these are from the time of Ptolemy and have been recorded by him. The times for these are explicitly stated to be in equal hours.

The third group consists of the single partial solar eclipse that has been called the eclipse of Theon.

152

Fotheringham [1909b] has used the magnitudes listed in Table V. 5. Newcomb [1875] has used the times but not the magnitudes. Fotheringham [1920] quoted results from an earlier paper by himself (Monthly Not. Roy. Astro. Soc. 80, 578-581, 1920) that used the times; I have not consulted this paper directly.

Although the eclipse records in Table V. 5 cover a span of more than five centuries, longer than the span of the other tables, I shall use a common estimate of the timing accuracy for all of them. It will be assumed that the timing error proper is one half that attributed to the Babylonian times in Table V. 2. That is, the standard deviation of the clock error will be taken to be 0. 05 times the interval from sunset or sunrise, whichever is nearer. In addition, there is the rounding error in reporting the times. Various fractions of an hour occur in Table V. 5. It is a reasonable presumption that the precision aimed at was about 1/6 hr. or 10^m. The standard deviation of the rounding error will thus be taken to be 3^m.

The records will receive comment, if only to list the chapter in Ptolemy [ca 152] where the record is found.

-200 Sep 22 M. Reference: Ptolemy; IV. 10. The magnitude was not given in the record. The eclipse had a magnitude of 8. 8 digits according to Oppolzer's Canon, so it is doubtful that the eclipse was total. (See the discussion of the record listed as -309 Aug 15b M (Hipparchus).) According to the record as preserved, the lunar eclipse began half an hour before the moon rose. This cannot possibly be an observation. It is probably the result of a computation connecting the length of an eclipse and its magnitude. If so, we do not know the extent, if any, to which it represents an observation independent of the end time. Hence this time is best ignored.

-199 Mar 19 M. Reference: Ptolemy; IV. 10.

-199 Sep 11 M. Reference: Ptolemy; IV. 10.

-173 Apr 30 M. Reference: Ptolemy; VI. 5. The times of the beginning, middle (defined as the half of the time), and the end were all listed. Since the middle time is exactly the average of the other times, it is probably not an independent observation and will be ignored.

-140 Jan 27 M. Reference: Ptolemy; VI. 5.

+125 Apr 5 M. Reference: Ptolemy; IV. 8.

133 May 6 M. Reference: Ptolemy; IV. 5.

134 Oct 20 M. Reference: Ptolemy; IV. 5.

136 Mar 5 M. Reference: Ptolemy; IV. 5.

364 Jun 16 M (Theon). Reference: Fotheringham [1920]. Theon, who was apparently an Alexandrian astronomer of the fourth century, left a record of the times connected with this solar eclipse in a commentary on Ptolemy's work. The eclipse was observed at Alexandria and was almost surely partial there. Times are given for the beginning, middle, and end phases; however, the time given for the middle phase differs by 8^{m} from being "half of the time". Since it is not quite clear what data are independent, all three times will be used but will be given the weight of only two observations.

Fotheringham assumed that the time of the phase called "middle" was the time of greatest phase. The discrepancy between the middle phase and the other two was reduced from 8^{m} to 5^{m} by this assumption. It is possible that the time of greatest phase was meant here, and that this time was estimated by estimating when the darkening was greatest.

CONJUNCTIONS AND OCCULTATIONS

1.   STARS APPEARING IN THE RECORDS OF CONJUNC-TIONS AND OCCULTATIONS

The records of conjunctions and occultations that will be used in this work are those preserved by Ptolemy [ca 152] and by Ebn Iounis [1008].   The records involve the moon, planets, and stars.   It is necessary to identify the stars and to establish the basis for finding their positions.

The Pleiades occur in the records in phrases such as "the eastern third of the Pleiades", but not with individual names.   Other stars that appear are identified in both sources by the descriptions used in Ptolemy's star table [Ptolemy, ca 152].   "The most southerly of the stars on the front of the Scorpion" is a typical identification in the conjunction and occultation records.   In Ptolemy's

TABLE VI. 1

STARS USED IN RECORDS OF CONJUNCTIONS AND OCCULTATIONS
(Epoch, Equinox, and Equator of 1950. 0)

Description	Name	Rt. Asc. (deg)	Declination (deg)	Proper Motion (deg/millenium)	
				Rt. Asc.	Decl.
Spica	67α Vir	200. 638 754	-10. 900 933	-0. 012	-0. 009
Most northerly on the front of the Scorpion	8β^1 Sco	240. 631 279	-19. 670 125	-0. 002	-0. 006
Middle star on the front of the Scorpion	7δ Sco	239. 342 992	-22. 480 936	-0. 003	-0. 007
Most southerly on the front of the Scorpion	6π Sco	238. 955 421	-25. 971 706	-0. 003	-0. 008
Heart of the Scorpion	21α Sco	246. 584 192	-26. 322 764	-0. 002	-0. 006
Heart of the Lion	32α Leo	151. 427 696	12. 212 372	-0. 071	+0. 001
Pleiades	25η Tauri	56. 126 767	23. 952 103	+0. 006	-0. 012
	17 Tauri	55. 475 229	23. 957 728	+0. 006	-0. 012
	19 q Tauri	55. 556 550	24. 311 920	+0. 006	-0. 012
	20 Tauri	55. 711 484	24. 213 058	+0. 006	-0. 012
	23 Tauri	55. 838 313	23. 794 164	+0. 007	-0. 012
	27 Tauri	56. 545 959	23. 902 128	+0. 005	-0. 012
	28 BU Tauri	56. 551 629	23. 985 433	+0. 004	-0. 013

table, the description is followed by the coordinates, which assist in the identification. Some stars in the table cannot be identified, probably because their coordinates have been garbled too badly in copying. Luckily, there is no apparent difficulty with the stars used in the conjunctions and occultations.

The stars involved are listed in Table VI.1. As a precaution, I identified them before consulting previous identifications. I have not found "heart of the Scorpion" and "heart of the Lion" in other analyses. The remaining stars in Table VI.1 appear in [Fotheringham, 1915a], and the identifications are identical. The coordinates and proper motions are taken from [Smithsonian Astro. Obser., 1966], with conversion of units. The names are those used in Becvar [1964]; the naming system used there is an elaboration of older systems and provides a link with older names.

2. CONJUNCTIONS AND OCCULTATIONS PRESERVED BY PTOLEMY

Ptolemy [ca 152, VII.3] has preserved records of seven conjunctions or occultations involving the moon and various stars. The stars are among those listed in Table VI.1.

Newcomb [1875] did not use these data. He feared that they formed a biased sample selected by Ptolemy to give credence to his wrong value for the precession of the equinox. Skepticism about data that accompany a bad conclusion is justified in general. In this case, because of other errors, Ptolemy did not need biased data to support his bad value of the precession. The errors in his values for the mean motions of the sun and moon are close (see Section II.2) to each other and to his error in the precession. Thus an unbiased sample of conjunctions and occultations would support Ptolemy's values[†]. Because of this,

---

[†]Superficially this looks like the argument used by Fotheringham [1915a] to answer Newcomb's objection, but it is actually quite different. Fotheringham upheld the validity

and because Ptolemy's transmittal of Hipparchus' equinox data seems valid (Section II. 2), I accept Ptolemy's transmittal of the conjunction and occultation data.

Fotheringham [1915a] gave translations of the relevant passages from Ptolemy. Each passage starts with a lengthy identification of the observer, the place, and the date. This is followed by a statement of the hour and then by a statement describing the circumstances. The statements of hour and circumstance, preceded by the date converted to the Julian calendar, are:

-294 Dec 21: ". . at the very beginning of the 10th hour, the Moon was observed to have overtaken with her north cusp the most northerly of the stars on the front of the Scorpion. "

-293 Mar 9: ". . as the 3rd hour was beginning, the Moon, with the middle of her disk on the exact east, overtook Spica, and Spica passed through her, cutting off one-third of her diameter exactly, on the north side. "

-282 Jan 29: ". . as the 3rd hour was ending, the southern half of the Moon was observed to have covered the eastern third part or half of the Pleiades exactly. "

-282 Nov 9: ". . when the 10th hour had lasted half an hour, when the Moon had risen from the horizon, Spica was observed exactly touching her north limb. "

+92 Nov 29: ". . as the 3rd hour of the night was beginning, the Moon with her south cusp occulted the southeast part of the Pleiades. "

98 Jan 11: ". . when the 10th hour had been completed, Spica was observed to be in occultation by the Moon,

---

of Ptolemy's "observations" and laid the blame for his bad conclusions upon the alleged badness of Hipparchus' observations; I do the opposite. Fotheringham [1918] later retracted his allegations about Hipparchus' data (Section V. 2 of this study).

for it was invisible; but as the 11th hour was ending, it was observed preceding the centre of the Moon by a distance of less than the Moon's diameter, being equidistant from the two cusps. " This constitutes two separate observations, which will be identified as 98 Jan 11a and 98 Jan 11b.

98 Jan 14: " . . as the 11th hour was ending, the south cusp of the Moon was seen in a straight line with the middle and the most southerly of the stars on the front of the Scorpion, but her centre was to the east of the straight line and as far distant from the middle star as the middle star was from the southern, and she appeared to have occulted the northernmost of the stars on the front for it was nowhere to be seen. "

TABLE VI. 2

CONJUNCTIONS AND OCCULTATIONS PRESERVED BY PTOLEMY

Date	GMT		Place	Interpretation Factor	Event
	h	m			
-294 Dec 21	01	27	Alex.	1	Conjunction, north cusp of moon with $8\beta^1$ Sco
-293 Mar 9	18	07	Alex.	1	Moon began to occult 67α Vir
-282 Jan 29	18	53	Alex.	1	Moon covered eastern third or half of Pleiades
-282 Nov 9	01	36	Alex.	1	Conjunction, north cusp of moon with 67α Vir
+ 92 Nov 29	17	00	Bithynia	1/3	Moon occulted southeast part of Pleiades
98 Jan 11a	04	11	Rome	1	67α Vir invisible
98 Jan 11b	05	25	Rome	1	67α Vir west of center of moon by less than lunar diameter
98 Jan 14	05	23	Rome	1/3	South cusp of moon on line with 7δ Sco and 6π Sco; $8\beta^1$ Sco occulted

The dates and times of these observations, converted to Greenwich Mean Solar Time, are listed in the first two columns of Table VI. 2. The place of observation appears in the third column. The circumstances described in the records are summarized in the last column. The column headed "Interpretation Factor" is a factor that will be used as "Reliability" was used with the large solar

158

eclipses in Chapter IV; that is, it will be used as a factor in computing the weight to be given each observation, in addition to the factors that arise from the estimated errors in observation. The interpretation factor represents my assessment of the confidence with which one can interpret the circumstances described in the records. The basis for assigning values to the factor will appear below in the discussions of individual records.

The dates of the Alexandria observations are relevant to the identification of the "eclipse of Hipparchus", since they create a strong presumption that Hipparchus had access to Alexandrian observations made soon after the founding of the city (Section IV. 5).

In the analysis of the records, the following coordinates will be used for the places of observation: Alexandria, 31°. 2N and 29°. 8E; Bithynia, 40°. 6N and 29°. 8E; and Rome, 41°. 9N and 12°. 8E.

In assigning times to the observations, it was assumed that the times are in unequal hours (see Section II. 3) after sunset at the place of observation. It was assumed that the times were even hours or half hours. Thus, "as the 3rd hour was beginning" was equated to 02^h, "as the 3rd hour was ending" was equated to 03^h, and "when the 10th hour had lasted half an hour" was equated to 09^h 30^m, all in unequal hours.

Fotheringham [1915a] assumed that the records implied that the hour was divided into thirds. Thus, he took "as the 3rd hour was beginning" to mean some time in the first 20 minutes of the 3rd hour and hence to mean $02^h 10^m$ on the average. Similarly, he replaced "as the 3rd hour was ending" by $02^h 50^m$. However, he used the even hour when a strong modifier accompanied "beginning" or "ending". This is a plausible interpretation. However, the means of the times assigned in the two ways differ negligibly. Inspection of the analysis in Fotheringham [1915a] shows that the times used in Table VI. 2 will lead to a smaller scatter in the residuals, but not by a significant

159

amount. Aside from the differences introduced by the different readings, there are small differences between the times in Table VI. 2 and those assigned by Fotheringham. These probably come from procedural differences in the reduction and are negligible.

I shall use the a priori estimate of the timing errors that was used in Section V. 5. There will also be an observational error coming from the range of lunar positions allowed by the description of the circumstances. For example, if all that was stated is that a certain star was occulted, the position of the moon can range over its diameter. I shall use the range divided by $2/3$ as the standard deviation of this error.

Fotheringham [1915a] assumed that a star would appear occulted to the observer if it were within $3'$ of arc of the bright edge of the moon, and he used this in figuring the range of lunar position allowed by the observation. He says that he did this (p. 388) "at Dr. Pickering's suggestion"; since he gives no source, Dr. Pickering presumably made the suggestion in a private communication. I have seen no data on the value that should be used, and since Dr. Pickering's suggestion sounds plausible, I shall adopt it. The value presumably depends upon the magnitude of the star, and $3'$ was probably intended to be typical of the stars in the records.

Some of the records need individual discussion.

-294 Dec 21 and 98 Jan 14: These records, and only these, involve $8\beta^1$ Sco, the most northerly of the stars on the front of the Scorpion. The observations with $8\beta^1$ Sco give rise to the two most discordant results in the analysis made by Fotheringham [1915a]. The discord in the record of -294 Dec 21 can be removed by postulating an error of 1^h in the time, but not the record of 98 Jan 14. It does not take a numerical analysis to show the trouble with the latter record. If the reader will plot the positions of the stars involved, as taken from Table VI.1, he will see that the distance from $8\beta^1$ Sco to the (extension of the) line joining 7δ Sco and 6π Sco is about two lunar diameters. Thus no part of the moon could be on this line when the moon is occulting 8β^1 Sco. Since it appears from Fotheringham's analysis that the moon was slightly more than half full with the bright side toward 8β^1 Sco, the discord in the observation is all the greater. This observation can be explained by postulating bad eyesight. In addition to bad eyesight, it requires a timing error of about 1^h in the same sense as the record of -294 Dec 21.

Both records can be explained by assuming that the star was about ½° west of the position calculated from Table VI.1. This means either that the present position is in error by about 0°.5 in right ascension or that the proper motion is wrong by about 0°.3 per millenium (about 1$''$ per year). Neither possibility seems at all likely, but it is curious that both these apparently accurate ancient observations can be reconciled by the same unlikely postulate.

8β^1 Sco has an apparent visual magnitude of 2.9 and is brighter than any other star in the vicinity. It forms a prominent triangle with 9ω^1 Sco and 14 ν Sco. It is unlikely that both observers would overlook it or identify it wrongly. Is it possible that there was once a star just west of the present 8β^1 Sco, which has now vanished but which was what the ancients meant by "the most northerly on the front of the Scorpion"? This too seems unlikely.

Since the trouble with the record of -294 Dec 21 is revealed only by computation, and since the record is quite clear, I shall give it an interpretation factor of 1. I shall use a different interpretation of the record of 98 Jan 14 from the one that Fotheringham [1915a] used. He put all the weight upon lining up the cusp with 7δ Sco and 6π Sco, and ignored the stated occultation. It seems harder to me to overlook the presence of 8β^1 Sco than to make a mistake about whether the point of a lunar cusp is on the extension of a line in the heavens. In the analysis in Part II, I shall place the moon where its distance from 8β^1 Sco (after allowing 3$'$ because the star is on the bright side) is half

161

the distance from its cusp to the imaginary line. Since the record cannot be interpreted at face value, I shall give it an interpretation factor of $\frac{1}{3}$.

-282 Jan 29: Fotheringham took "the eastern third part or half of the Pleiades exactly" to mean that the three most easterly of the Pleiades, namely 25η Tauri, 27 Tauri, and 28 BU Tauri, were occulted. I am not clear whether he did this on the assumption that "third part" referred to extent and "half" to the number of stars or not. I would translate "επομενον ητοι γ η ζ' μερος" as "eastern third or half part" rather than put "part" where Fotheringham did. There is also the problem that the observer Timocharis used "ακριβος" or "exactly". It seems odd to say "a third or a half exactly". The only explanation I have thought of is to assume that Timocharis (or a later editor) wanted to say something more accurate than would either "half" or "third" by itself, that is, that he meant to say that the moon covered more than a third and less than a half of the Pleiades. This reading sounds to me a little more plausible than Fotheringham's, but not much.

I shall take the record to mean that the western edge of the moon was east of the center of the Pleiades by 1/12 of the angular extent of the Pleiades, with a total range in position of 1/6 of their angular extent. The limits of position assigned to the moon are about the same on either interpretation. Thus the interpretation factor will be taken as unity in spite of the uncertainty in the meaning of the language.

92 Nov 29: Fotheringham [1915a] pointed out a difficulty about this record that was revealed by his computations. He found that the moon on this occasion passed so far north that it could not occult the southeast part of the Pleiades. He assumed that the observer Agrippa meant to say "northwest", and used the record on that assumption. A copying error in which a copyist replaces a term by its opposite is plausible in many cases, but it seems unlikely here since "southeast" and "northwest" would not have been written as single words. The preserved record has

"επομενον και νοτιον" (literally "following and southern").
Fotheringham's assumption would require the original
record to read "προηγουμενον και αρκτον" (preceding and
northern) if the usual terminology had been followed. The
postulated error requires changing two words simultaneously[†].

Another interpretation is that Agrippa originally
wrote something equivalent to saying that the moon was
over the southeastern Pleiades, meaning that it was upward
from them as he viewed them; this would put the moon to
the north of them. A copier or editor might have thought
he meant "επεκαλυψε" , which means that it was over
them in the sense of hiding them, and substituted this word
for the original. This word is a standard term used to
describe an occultation. I shall adopt this interpretation,
assuming that Agrippa meant that the moon was "over" or
north of the southeastern Pleiades, and use an interpretation factor of $\frac{1}{3}$. In the unlikely event that Fotheringham's
calculation was seriously in error and that the moon could
have occulted the southeastern Pleiades, I shall change the
factor to unity in Part II.

There are no apparent problems with the remaining
records and they will receive interpretation factors of
unity. It will be assumed that the clock errors in the observations called 98 Jan 11a and 98 Jan 11b are correlated,
and the final weights of these will be adjusted accordingly.
That is, these are not quite independent observations.

The reader may object that the comments on the
records violate the spirit of Part I, which is not supposed
to be biased in its textual interpretation by knowledge of
computed results. However, he should note that interpretation factors other than unity were assigned only when an

---

[†]The important question is perhaps how many ideas were
involved, not how many words were involved. Did the hy-
pothetical copier regard "preceding and northern" as a single notion whose opposite was the single notion "following
and southern", or did he consider that these phrases em-
bodied two separate notions each?

earlier computation had revealed a physical impossibility on the basis of the obvious interpretation. I have merely decided in Part I rather than in Part II on a tentative way to deal with the apparent impossibilities, which would have to be faced at some point in the work.

3. CONJUNCTIONS AND OCCULTATIONS PRESERVED BY EBN IOUNIS

Ebn Iounis [1008] preserved the records of a number of conjunctions and occultations involving the planets. The outer planets have such small mean motions that conjunctions or occultations involving only them, or one of them and a star, do not define an epoch with precision. Hence such conjunctions or occultations are not useful for the purposes of this study and will be omitted. The remaining events involve Venus and another body; in a few cases, the other body is Mercury.

The records in which Venus occurs will next be presented, along with comments where needed. The records are found on pages 94 to 112, 156 to 158, and 180 to 210 in the edition of Ebn Iounis cited. They do not appear in strictly chronological order in the reference. A few records that do not seem to give enough data will be omitted. The date of a record will be given first, followed by the place of observation, then a quotation, and finally the comments, if any.

858 Aug 28, Bagdad. "I saw Venus and Saturn . . . about (or toward) dawn. Venus still had about 2/5 of a degree to go before reaching Saturn, . . . Venus was a little north of Saturn . . ." Sunrise will be used as the time for this observation, even though the language implies that the time was slightly earlier. If sunset is also used as the time for other events that occurred slightly after sunset, the errors made by such usage will tend to cancel.

864 Feb 13, Bagdad. "Mars and Venus . . seemed, to the sight, to touch at the beginning of the night . . ."

The time of this event will be taken as sunset; as said in the preceding paragraph, the resulting error tends to cancel the error probably made there.

864 Oct 10, Bagdad. "I saw . . at dawn . . Venus and Mars making, so to speak, only a single planet; . ."

885 Sep 9, Bagdad or Shiraz. "I saw . . a perfect occultation of the heart of the lion[†] by Venus, the morning of the 6th férie[‡], . . . an hour before sunrise. Venus was, on the morning of the 5th férie, more than a degree from the heart of the lion; the morning of the 7th férie, she was more advanced by the same quantity; and the morning of the 6th férie, the heart of the lion was not visible." This careful observation would probably be quite valuable if anyone had remembered to tell us where it was made.

We are told [Ebn Iounis, 1008, p. 158] that the record is from the observations of Aboulhassan Ali ebn Amajour al Turki. We are told on page 104 of the same reference that he was one of two astronomers Amajour, father and son. We are told on page 110 that the father observed for 30 years and perhaps longer. The last report attributed to an Amajour is the eclipse record 933 Nov 5 Is, Section V. 4. Some of the reports are attributed to Ali ebn Amajour and some to Aboulhassan Ali ebn Amajour; in some of the former, the recorder refers to "my son Aboulhassan" as a participant. All this leaves it unclear whether the son already made the observation on 885 Sep 9[*], whether both men had the same names, or perhaps whether the son was the transmitter and not the observer of the 885 Sep 9 record.

[†] 32α Leo; see Table VI. 1. Ebn Iounis gives the longitude of the star he calls "heart of the lion" at several epochs. The close agreement between the longitudes given and those calculated from modern tables puts the identification beyond serious question.

[‡] That is, the 6th day of the week.

[*] If he did, he had a professional life of at least 48 years.

Many of the Amajour observations beginning with the eclipse 923 Jun 1 Is (Section V. 4) are located at Bagdad. The others after this date are not specifically located but can be placed at Bagdad with reasonable confidence. Before this date, only the observation of 901 Oct 4, to be given later in this section, is specifically located; it is placed at Shiraz.

Two possibilities suggest themselves. Either the Amajours observed at Bagdad all their working lives but one of them happened to be at Shiraz on 901 Oct 4, or they observed first at Shiraz and moved to Bagdad between 901 and 923. There are other possibilities, of course, but these seem superficially most probable. The evidence that I have seen does not suggest a choice, so I shall give these possibilities equal weight. The observation of 885 Sep 9 will be analyzed as if it had been made at the point midway between Bagdad and Shiraz.

901 Oct 4, Shiraz. "I saw, . . at Shiraz, Venus in conjunction with Jupiter . . . at dawn . . There was between the two planets the interval of a fetr[†] . . Venus was to the north, and it seemed to me that she reached Jupiter at sunrise, . . ." This is an Amajour observation.

918 Dec 24, Bagdad or Shiraz. "I saw . . Venus with the heart of the scorpion[‡], which was then at 24° 31' of the scorpion[‡]; and I found Venus at 29° of the scorpion, . . ." The observer Amajour goes on to say that the time was $5^h 50^m$, eq. hours, before noon.

On first reading, this report sounds garbled. It seems to say as a single observation that Venus was in conjunction with 21α Sco and that it was several degrees away. It probably is not garbled, because several Amajour

---

[†] The translator believed that a fetr was about 40'.

[‡] The star is 21α Sco (Table VI. 1), and its longitude at the epoch of the observation is stated to be 210° + 24° 31' = 234° 31'.

observations involving other planets have similar forms. I shall assume that Amajour meant that he located Venus with (the aid of) 21α Sco, and that Venus was 29° - 24° 31′ = 4° 29′ east of 21α Sco.

I shall use the point midway between Bagdad and Shiraz in analyzing this observation.

There are a number of other Amajour observations, in which both the place and the hour are missing. The probable reason for this is that Ebn Iounis used many Amajour observations to show that existing planetary tables were in error by several degrees. The place and hour were not needed for this purpose; the day was sufficient.

The remaining observations in this section were made by Ebn Iounis himself.

987 Jun 17, Cairo. "Conjunction of Venus and the heart of the lion, at sunset: it should have happened at 8 eq. hours after noon . . ." The exact time given appears to be the result of a calculation. It is not clear whether the calculation was based upon observations or upon a set of tables. However, in an earlier record Ebn Iounis used a similar construction in a case where it seems likely that he based the calculation upon his observations. Hence, I shall assume that 8^h is the time of the conjunction and that "sunset" was used merely as a general indication of the circumstances.

988 Jan 20, Cairo. "Conjunction of Saturn and Venus in the first degree of Capricorn. I saw Venus and Saturn in conjunction the day of the 6th férie, about half an hour before sunrise. There was about a digit[†] of latitude between them. Venus was north of Saturn. Jupiter was near these planets and preceded[‡] them by about 1°." This must have been a spectacle.

[†] Probably 1/12 of a degree.

[‡] Was to the west of.

990 Jun 22, Cairo. "Conjunction of Venus and the heart of the lion in the west. I saw them in conjunction the 2nd férie, about an hour after sunset. Venus was to the north of the star. There was about a shebr[†] between them in latitude. I observed them for several days, and I have no doubt about the certainty of my observation."

991 Dec 22, Cairo. "Conjunction of Venus and Saturn . . . about 6 eq. hours after noon . . Saturn was north of Venus. Their difference in latitude, 1° or a little more, according to my estimate; their distance, about 1 shebr and two knots[‡]." Since both a latitude difference and a "distance" are given and since they are about the same size, it is not clear if the distance is a repetition of the latitude difference in different terms or if it is a measured separation in longitude. Because of this ambiguity, and because the unit used for the distance is unknown, I shall not use this observation. I shall calculate the circumstances in order to see whether the meaning of the passage can be inferred.

992 Sep 16, Cairo. "Conjunction of Venus and the heart of the lion in the east . . . an equal hour before sunrise. Venus had already passed the heart of the lion by about a third of a degree; she was to the south. Difference in latitude, about a half degree."

995 Jan 3, Cairo. "Conjunction of Venus and Mercury in the west . . . after sunset.[*] They were separated by about 1 shebr[**]. Venus preceded Mercury. Height of

---

[†] The translator cited believed that a shebr was about 1°.

[‡] The cited translation has the French word noeuds. The translator comments that he has not seen this term elsewhere. He speculates that it may mean an ancient unit equal to 1/9 of a degree.

[*] A translator's note says that the Arabic term used here means the end of twilight. For reasons given in connection with the record of 858 Aug 28, I shall use sunset anyway.

[**] I shall take this to be 1°.

168

Venus 10°. They followed the same route and I believe that Mercury eclipsed Venus when he reached her." "Preceded" usually means "to the west of", but the subsequent wording here suggests that perhaps the opposite was meant. If the calculations cannot resolve the ambiguity, this observation will not be used.

995 Jun 11, Cairo. "Conjunction of Jupiter and Venus . . at 7 eq. hours after noon . . They were between the meridian and the setting, Venus to the north of Jupiter; their distance in latitude about a fetr. The pole of the ecliptic was between the meridian and the east. It was highly elevated, and Venus for that reason should appear above Jupiter at the time of conjunction: she was effectively more elevated when the great circle which passes through the poles of the ecliptic seemed to pass through the centers of both planets at the same time." This is the first record that gives this much detail and in which it is quite specific that the conjunction was judged in longitude rather than in right ascension. The translator, in a note, expresses the opinion that the passage implies the use of an armillary sphere or some such instrument.

995 Jun 18, Cairo. "Conjunction of Venus and the heart of the lion. It arrived at the _____ setting[†], at about $7^h 40^m$, eq. hours, after noon . . . Venus was to the north of the heart of the lion; their distance in latitude, 2/3 or 3/4 degree, about a fetr to the sight." Ebn Iounis ends this report in a manner similar to the preceding one. Whether the repetition of the distance in latitude by means of the fetr is merely intended to give the measurement in other units, or whether a measurement "to the sight" is a measurement of a different quantity, such as the distance in declination, is uncertain.

[†] The French has couchant du méridien. Since I do not see how to interpret either "setting of the south" or "setting of the meridian", I have left a blank. Comparison with the wording of the preceding record suggests that there may have been a copying error.

996 Aug 8, Cairo. "Conjunction of Jupiter and Venus
. . about a half-hour after sunset . . Venus was to the north
of Jupiter, which seemed to touch her[†]. I evaluated their
distance in latitude at about 5'."

997 May 24, Cairo. "Venus eclipsed Saturn in a
manner not to be doubted, about 2/3 of an equal hour before
sunrise . . ."

998 June 4, Cairo. "Conjunction of Mars and Venus.
I saw them in conjunction at the beginning of the night . .
at 8 eq. hours, about, after noon of the second férie.
There was about a digit between them, to the eye, or a
little less. Venus was north of Mars and more elevated on
the horizon. The great circle passing through the poles of
the ecliptic and by Venus showed that she had already
passed Mars by about 1/4 of a degree." If "digit" does
mean 1/12 of a degree, the observations seem inconsistent.
However, Ebn Iounis may merely mean that a digit was his
estimate made before he went to the observing instrument,
or, more likely, that a digit was the estimated separation
in latitude. I shall assume that the observation means that
Venus was 15' east of Mars in longitude.

998 Jun 23, Cairo. "Conjunction of Venus and the
heart of the lion . . about an hour after sunset, 8 eq. hours
after noon . . Venus was to the north, about 1° away in
latitude."

999 Apr 9, Cairo. "I saw Venus at the end of the
night before the 2nd férie, $16^h 30^m$ eq. hours, after noon
of the 1st férie . . She was with Mars in the east, preceded him by about 1°, and followed the same route."

1000 May 19, Cairo. "Conjunction of Mercury and
Venus . . . about 8 eq. hours after noon . . . Mercury was
north of Venus. Their difference in latitude, 1/3 degree."

[†]The translator comments that the clause about touching
has been left blank in some copies of Ebn Iounis, but that
he thinks he has deciphered it.

170

1001 Jun 2, Cairo. "Conjunction of Venus and Mercury in Cancer. I saw them 1 eq. hour after sunset. Venus was north of Mercury and a little below it. Mercury was very hard to see. I determined their conjunction at midnight . . ." Although Ebn Iounis does not say so explicitly, he presumably measured the separation in longitude 1^h after sunset and calculated the time of conjunction from the measurement.

1002 Apr 18, Cairo. "Conjunction of Jupiter and Venus in the Fishes. I saw them, the 7th férie, about 1^h 30^m, eq. hours, before sunrise. Venus was still 1/5 degree away from Jupiter . . . I determined their conjunction at 2 eq. hours before noon . . The conjunction took place in longitude and in latitude."

1002 Jul 14, Cairo. "Conjunction of Venus and Saturn . . . about 8 eq. hours after noon of the 3rd férie. In fact, I observed them the 3rd férie, an hour and a little before sunrise. Venus had not yet reached Saturn. I observed them the 4th férie at the same hour, . . . and she was more easterly by about 1/3 degree."

1003 Jan 7, Cairo. "Conjunction of Venus and Mars . . When I observed them, there was a fetr, about a half-degree[†], between them in latitude. I estimated that she had passed him by about a half-degree, and I determined their conjunction at 12 eq. hours after noon . . ."

1003 Feb 18, Cairo. "Conjunction of Jupiter and Venus . . I observed them about 1/3 hour after sunset on the 5th férie. Jupiter preceded Venus by about 1/3 degree, and was a little to the north . . I observed them again the 6th férie, about 1/3 hour after sunset. Venus had already passed Jupiter by 2/3 degree. I determined their conjunction at 14 eq. hours after noon of the 5th férie. "

[†]This does not seem to agree with the earlier estimate that the fetr was about 40'. However, Ebn Iounis may not have intended this statement to be as precise as the difference between 30' and 40'.

1003 Jun 18, Cairo. "Conjunction of Venus and the heart of the lion in the west. I observed them after sunset, . . difference in latitude about 1/4 degree. Venus had only a little way to go to reach the heart of the lion. I determined their conjunction at 14 eq. hours after noon . . ."

This ends Ebn Iounis' records of conjunction and occultations.

When a single observation and a calculated time of conjunction are given, the observation will be used in preference to the calculated time. Rarely, two observations and a calculated time are given. In such a case, one could use the two observations independently. However, it is likely that there are systematic errors that would be

TABLE VI. 3

ISLAMIC RECORDS OF CONJUNCTIONS OR OCCULTATIONS
INVOLVING VENUS

Date	Place	Local Time	Second Body	Longitude of Venus Relative to Second Body
858 Aug 28	Bagdad	SR	Saturn	0°. 4W
864 Feb 13	Bagdad	SS	Mars	0
864 Oct 10	Bagdad	SR	Mars	0
885 Sep 9	a	SR-1^h	32α Leo	0
901 Oct 4	Shiraz	SR	Jupiter	0
918 Dec 24	a	N-5^{h}50^{m}eh	21α Sco	4°29'E
987 Jun 17	Cairo	N+8^{h}eh	32α Leo	0
988 Jan 20	Cairo	SR-1/2^h	Saturn	0
990 Jun 22	Cairo	SS+1^h	32α Leo	0
991 Dec 22^b	Cairo	N+6^{h}eh	Saturn	?
992 Sep 16	Cairo	SR-1^{h}eh	32α Leo	20'E
995 Jan 3	Cairo	SS	Mercury	1°W
995 Jun 11	Cairo	N+7^{h}eh	Jupiter	0
995 Jun 18	Cairo	N+7^{h}40^{m}eh	32α Leo	0
996 Aug 8	Cairo	SS + 1/2^h	Jupiter	0
997 May 24	Cairo	SR-2/3^{h}eh	Saturn	0
998 Jun 4	Cairo	N+8^{h}eh	Mars	15'E
998 Jun 23	Cairo	N+8^{h}eh	32α Leo	0
999 Apr 9	Cairo	N+16^{h}30^{m}eh	Mars	1°W
1000 May 19	Cairo	N+8^{h}eh	Mercury	0
1001 Jun 2	Cairo	M	Mercury	0
1002 Apr 18	Cairo	SR-1^{h}30^{m}eh	Jupiter	0°. 2W
1002 Jul 15	Cairo	SR-1^h	Saturn	20'E
1003 Jan 7	Cairo	N+12^{h}eh	Mars	0
1003 Feb 18	Cairo	N+14^{h}eh	Jupiter	0^c
1003 Jun 18	Cairo	N+14^{h}eh	32α Leo	0

^{a}The point midway between Bagdad and Shiraz will be used.

^{b}The record is ambiguous and will not be used in the final inference.

^{c}Will receive extra weight. See main text.

172

common to the two measurements, and treating the observations independently would give them too much weight. In such a case, I shall use the calculated time, and give it $\sqrt{2}$ times the weight of a single observation. This choice is arbitrary.

The records are summarized in Table VI. 3. The first two columns are self-explanatory. The third column gives the local time as it was recorded. In this column SR, SS, N, and M mean the times of sunrise, sunset, noon, and midnight, respectively. When the time interval from one of these epochs was explicitly stated to be given in equal hours, this is indicated by putting "eh" after the time. The fourth column gives the second body involved in the observation and the fifth column gives the observed longitude of Venus relative to this body.

Because three different conventions about the hour that begins a day are involved in finding the dates given in Table VI. 3, we may expect more errors in the calendar date than usual. Except when the daily motion of Venus is small, we can remove an ambiguity of an integral number of days by the calculations to be performed. If the motion is too slow to remove a dating ambiguity, the observation is useless for our purposes even if the date is correct.

When the time is not explicitly stated to be in equal hours, it will be assumed that unequal hours were used.

The coordinates that will be used for the places of observation are: Bagdad, 33°. 3N, 44°. 4E; Cairo, 30°. 0N, 31°. 2E; and Shiraz, 29°. 6N, 52°. 6E.

Most of the times are reported only to the nearest hour, although the time-measuring ability of the Islamic astronomers (Section V. 4) was minutes rather than hours. The probable reason is that the astronomers realized that their errors in measuring angles for a body moving as slowly as Venus dominated their time errors, and that the nearest hour was sufficient.

173

Inspection of the angular measurements given by the Islamic astronomers suggests that they felt that 5′ was about the limit of their angle-measuring ability for a single reading, and this number sounds plausible. Since the average geocentric angular velocity of Venus is about 1° per day, the corresponding time error is about 2^h. Thus the observers were justified in rounding their times to an even hour.

The angular coordinates quoted in the records are ecliptic rather than equatorial coordinates; correspondingly, we may assume that conjunctions were estimated with regard to longitude rather than to right ascension even when this is not explicitly stated. Longitude is not an easy angle to measure, and it is plausible that the error in judging the longitude difference between two bodies increases as the latitude difference between them increases. Since there is no direct evidence about this dependence, I shall follow the simplest course. This is to use 5′ as the a priori estimate of the standard deviation in the longitude difference between two bodies with nearly the same latitude.

If the standard deviation in longitude difference is 5′, the standard deviation in finding a value of ephemeris time from the data is about 2^h for a single observation, and about $0^h.4$ for the set of about 25 observations. For the mean epoch of the observations, the difference between solar time and ephemeris time is of the order of $0^h.8$. Thus it is reasonable to expect that the Venus conjunctions will make a useful but not a dominant contribution toward finding the acceleration of the earth.

There is a possible source of error for which no a priori estimate can be given. In a number of records it is stated that a conjunction occurred even though subsequent statements concerning longitude measurements show explicitly that the observation was not of a conjunction. When the longitude measurement is given, loose use of the word "conjunction" does no harm. The possible error is that "conjunction" was used loosely without an accompanying (or with a now-lost) longitude measurement.

174

REFERENCES FOR PART I

This list is restricted to references that have been directly consulted in the course of preparing Part I of this study. Other works that come up in discussion, and that probably contain relevant information although they have not been directly consulted, are cited parenthetically or in footnotes.

Al-Battani, Astronomical Work, ca 925. There is a translation into Latin by Nallino, C. A., Publ. Reale Oss. di Brera, Milano, 1903.

Appleby, J. T. (ed. and trans.), The Chronicle of Richard of Devizes, Thom. Nelson and Sons, London, 1963.

Becvar, A., Atlas of the Heavens, Catalogue 1950. 0, Czechoslovak Acad. of Sciences, Prague, or Sky Publishing Corp., Cambridge, Mass., 1964.

Bede, Historia Ecclesiastica, 731. This paper uses the translation by J. Stevens (1723), reprinted by J. M. Dent and Sons, London, 1910.

Biot, M., Résumé de Chronologie Astronomique, Mem. Acad. Sci. Paris, 22, pp. 209-476, 1849.

Britton, J. P., "On the Quality of Solar and Lunar Observations and Parameters in Ptolemy's Almagest," a dissertation submitted to Yale University, 1967.

Celoria, G., "Sull 'Eclissi Solare Totale del 3 Giugno 1239," Memorie del Reale Istituto Lombardo di Scienze e Letteri, Classe di Scienze Matematiche e Naturali, 13, pp. 275-300, 1877a.

_____, "Sugle Eclissi Solari Totali del 3 Giugno 1239 e del 6 Ottobre 1241," Mem. del Reale Inst. Lomb. di Sci. e Lett., Classe di Sci, Mat. e Nat., 13, pp. 367-382, 1877b.

Chambers, G. F., The Story of Eclipses, Appleton and Co., New York, p. 107, 1915.

Cherniss, H., Introduction to De Facie Quae in Orbe Lunae Apparet, in Plutarch's Moralia, translated by Cherniss, H. and Helmbold, W. C., Harvard University Press, Cambridge, Mass., 1957.

Clemens, S. L., A Connecticut Yankee in King Arthur's Court, Harper and Bros., New York, 1889.

Clerke, A. M., "Astronomy: History," Encyclopedia Britannica (11th ed.), 2, Encyclopaedia Britannica, Inc., New York, 1910.

Curott, D. R., "Earth Deceleration from Ancient Solar Eclipses," Astron. J., 71, pp. 264-269, 1966.

Dandekar, B. S., "Measurements of the Zenith Sky Brightness and Color During the Total Solar Eclipse of 12 November 1966 at Quehua, Bolivia," Applied Optics, 7, pp. 705-710, 1968.

de Saussure, L., Les Origines de l'Astronomie Chinoise, Maissoneuve, Paris, 1930. A. Pogo, in a review of this book in Isis, 17, pp. 267-271, 1932, shows that the editor who collected de Saussure's papers into this volume did not change de Saussure's original page numbers in the cross references, and shows how to use the printed cross references.

de Sitter, W., "On the Secular Accelerations and the Fluctuations of the Longitude of the Moon, the Sun, Mercury and Venus," Bull. Astron. Inst. of the Netherlands, IV, pp. 21-38, 1927.

Dickins, B., "The Cult of St. Olave in the British Isles," Saga-Book of the Viking Society, XII, part II, pp. 53-80, 1940.

Dio Cassius, Romaike Istoria, ca 230. Translated in Cary, E., Dio's Roman History, Loeb Classical Library, Heinemann, London and MacMillan, New York, 1914.

Diodorus (Siculus), Historical Library, Bk. 20, ca -20. Reprinted, with English translation by R. M. Geer, Harvard University Press, Cambridge, Mass., 1954.

Dodgson, C. L., Through the Looking Glass, MacMillan and Co., London, 1872.

Dreyer, J. L. E., "Early Observations of the Solar Corona," Nature, 16, p. 549, 1877.

Dubs, H. H., The History of the Former Han Dynasty, a critical translation with annotations, Waverly Press, Baltimore, 1, 1938; 2, 1944; 3, 1955.

Duncombe, R. L., private communication, May 1968.

Ebn Iounis, (or Ibn Junis), Le Livre de la Grande Table Hakémite, 1008. There is a translation into French by le C. en Caussin, Imprimerie de la République, Paris, 1804.

Encyclopedia Brittanica, unsigned article "Roland, Legend of," (11th ed.), 23, Encyclopaedia Britannica, Inc., New York, 1911.

Eusebius Pamphili, Chronicon, ca 325. There is a collated edition of the principal texts edited by A. Schoene, Weidmann's, Berlin, 1866 and 1875.

Explanatory Supplement to The Astronomical Ephemeris and The American Ephemeris and Nautical Almanac, H. M. Stationery Office, London, 1961.

Fotheringham, J. K., "The Eclipse of Hipparchus," Monthly Not. Roy. Astro. Soc., 69, pp. 204-210, 1909a.

———, "On the Accuracy of the Alexandrian and Rhodian Eclipse Magnitudes," Monthly Not. Roy. Astro. Soc., 69, pp. 666-668, 1909b.

———, (assisted by Gertrude Longbottom), "The Secular Acceleration of the Moon's Mean Motion as Determined from the Occultations in the Almagest," Monthly Not. Roy. Astro. Soc., 75, pp. 377-394, 1915a.

———, "Note on Some Results of the New Determination of the Secular Acceleration of the Moon's Mean Motion," Monthly Not. Roy. Astro. Soc., 75, pp. 395-396, 1915b.

———, "The Secular Acceleration of the Sun as Determined from Hipparchus' Equinox Observations; with a Note on Ptolemy's False Equinox," Monthly Not. Roy. Astro. Soc., 78, pp. 406-423, 1918.

———, "A Solution of Ancient Eclipses of the Sun," Monthly Not. Roy. Astro. Soc., 81, pp. 104-126, 1920.

———, "Historical Eclipses," the Halley Lecture of May 17, 1921, in Oxford Lectures on History, Oxford Press, 1921.

———, "Eclipse: Ancient Eclipses," Encyclopedia Britannica, 7, Encyclopaedia Britannica, Inc., Chicago, 1958.

Frazer, Sir J. G., The Golden Bough (abridged edition), MacMillan Co., New York, 1922.

Gaubil, le Père, Une Histoire de l'Astronomie Chinoise, being 2 of Observations Mathématiques, etc., E. Souciet, Ed., Rollin Père, Paris, 1732a. (See Souciet, 1729.)

———, Un Traité de l'Astronomie Chinoise, being 3 of Observations Mathématiques, etc., E. Souciet, Ed., Rollin Père, Paris, 1732b. (See Souciet, 1729.)

Gilbert, Sir W. S., The Mikado, 1885. Reprinted in The Complete Plays of Gilbert and Sullivan, Random House, New York.

Ginzel, F. K., Spezieller Kanon der Sonnen- und Mondfinsternisse, Mayer and Müller, Berlin, 1899.

Hayden, Fr. F. J., private communication, April 16, 1968.

Herodotus, History, ca -446. There is a translation by George Rawlinson, first published in 1858 and reprinted by Tudor Publishing Co., New York, 1947.

Hoang, P., Catalogue des Éclipses de Soleil et de Lune (Variétés Sinologiques No. 56), Imprimerie de la Mission Catholique, Shanghai, 1925.

Holland, A. W., "Charlemagne," Encyclopedia Britannica, (11th ed.), 5, Encyclopaedia Britannica, Inc., New York, 1910.

Jebb, Sir R. C., "Thucydides," Encyclopedia Britannica, (11th ed.), 26, Encyclopaedia Britannica, Inc., New York, 1911.

Kratz, H., private communication, January 1969.

Kugler, F. X., Sternkunde und Sterndienst in Babel, Aschendorffsche Verlagsbuchhandlung, Münster, Westfalen (Germany), 1909.

Landmark, J. D., "Solmørket over Stiklestad," Kong. Norske Vidensk. Selsk. Skrifter for 1931, No. 3, pp. 1-62, 1931.

Macartney, C. A., "Hungary: History, the Kingdom to 1526," Encyclopedia Britannica, 11, Encyclopaedia Britannica, Inc., Chicago, 1958.

Munk, W. H. and MacDonald, G. J. F., The Rotation of the Earth, Cambridge University Press, Cambridge, 1960.

179

Naval Observatory (U.S.), Ancient Sun and Moon, unpublished tables, 1953. The preparation of the tables is described by Woolard, E. W., in "Theory of the Rotation of the Earth Around Its Center of Mass," Astro. Papers Prepared for the Use of the American Ephemeris and Nautical Almanac, XV, Part I, U.S. Government Printing Office, Washington, 1953.

Needham, J., Science and Civilization in China, 3, (with collaboration of Wang Ling), Cambridge University Press, Cambridge, 1959.

Neugebauer, P. V., Astronomische Chronologie, W. de Gruyter and Co., Berlin, 1929.

Neugebauer, O., "The Water Clock in Babylonian Astronomy," Isis, 37, pp. 37-43, 1947.

_____, Astronomical Cuneiform Texts, Lund Humphries, London, England, 1955.

_____, The Exact Sciences in Antiquity, Brown University Press, Providence, Rhode Island, 1957.

Newcomb, S., "Researches on the Motion of the Moon," Washington Observations, U.S. Naval Observatory, 1875.

_____, "Eclipse," Encyclopedia Britannica, (11th ed.), 8, Encyclopaedia Britannica, Inc., New York, 1910.

Oppolzer, T. R. von, "Note über eine von Archilochos erwähnte Sonnenfinsterniss," Sitzber, der k. Akad. der Wissenschaften, Wien, 86, Bd. 2, pp. 790-793, 1882.

_____, Canon der Finsternisse, Kaiserlich-Königlichen Hof- und Staatsdruckerei, Wien, 1887. Translation by O. Gingerich into English, printed by Dover Publishing Co., New York, 1962.

Pannekoek, A., A History of Astronomy, Interscience Publishing Co., New York, 1961.

Parker, R. E., private communications, April and May 1968.

Peters, C. H. F. and Knobel, E. G., Ptolemy's Catalogue of Stars, A Revision of the Almagest, Carnegie Institution of Washington, Publication No. 86, Washington, D. C., 1915.

Pliny the Younger (Gaius Plinius Caecilius Secundus), Letters, ca 100. There is a translation by William Melmoth, revised by F. C. T. Bosanquet, printed in Harvard Classics, 9, Collier and Son, New York, 1909.

Plummer, C., "Anglo-Saxon Chronicle," Encyclopedia Britannica, (11th ed.), 2, Encyclopaedia Britannica, Inc., New York, 1910.

Plutarch, De Facie Quae in Orbe Lunae Apparet, ca 90. This paper uses the translation by Harold Cherniss in Plutarch's Moralia, Cherniss, H. and Helmbold, W. C., 12, Harvard Press, Cambridge, Mass., 1957.

_____, Parallel Lives, ca 100. Dryden's translation is reprinted in part in Harvard Classics, 12, Collier and Son, New York, 1909.

Ptolemy, C., 'E Mathematike Sýntaxis (The Almagest), ca 152. This paper uses both the translation into French by M. Halma, Henri Grand Libraire, Paris, 1813, and the translation into German by K. Manitius, Leipzig (no publisher given), 1913. The chapter division and numbering is Halma's; there is an occasional difference of 1 from Manitius' numbering.

Sarton, G. , A History of Science; Ancient Science Through the Golden Age of Greece, Harvard University Press, Cambridge, Mass. , 1952.

Schroader, I. H. , private communications, April and May 1968.

Smith, S. , "Chronology: Babylonian and Assyrian, " Encyclopedia Britannica, 5, Encyclopaedia Britannica, Inc. , Chicago, 1958.

Smithsonian Astrophysical Observatory, Star Catalogue, Positions and Proper Motions of 258, 997 Stars for the Epoch and Equinox of 1950. 0, 1 through 4, Smithsonian Institution, Washington, D. C. , 1966.

Snorri (Sturluson), Noregs Konunga Sogur (Heimskringla), ca 1230. There is an edition edited by Finnur Jonsson, G. E. C. Gads Forlag, Kobenhavn, 1911. There is a translation into English by L. M. Hollander, University of Texas Press, Austin, Texas, 1964.

Souciet, E. , Ed. , Observations Mathématiques, Astronomiques, Géographiques, Chronométriques, et Physiques Tirées des Anciens Livres Chinois, 1, Rollin Pere, Paris, 1729.

Spencer Jones, H. , "The Rotation of the Earth, and the Secular Accelerations of the Sun, Moon, and Planets, " Monthly Not. Roy. Astro. Soc. , 99, pp. 541-558, 1939.

Storm, Gustav, Islandske Annaler indtil 1578, Grøndahl and Søns Bogtrykkeri, Christiana, 1888.

Thucydides, History of the Peloponnesian War, Bk. 2, 28, ca -420. This paper uses the translation by Richard Crawley, J. M. Dent and Sons, London, 1910.

Turoldus, La Chanson de Roland (Oxford manuscript), ca
1100. There is a translation into modern French
prose by Joseph Bédier, published by H. Piazza,
Paris; the 27th printing is dated 1922.

van der Waerden, B. L., Die Anfänge der Astronomie, P.
Noordhoff Ltd., Groningen, The Netherlands, (German ed.) 1956.

_____, "Drei umstrittene Mondfinsternisse bei Ptolemaios,"
Museum Helveticum, 15, fasc. 2, pp. 106-109, 1958.

_____, "Secular Terms and Fluctuations in the Motions of
the Sun and the Moon," Astron. J., 66, pp. 138-147,
1961.

Webster's Biographical Dictionary, G. & C. Merriam Co.,
Springfield, Massachusetts, 1953.

Weston, Jessie L., From Ritual to Romance, Cambridge
University Press, Cambridge, 1920. Reprinted by
Doubleday, Garden City, New York, 1957.

Whitelock, D. (ed.), The Anglo-Saxon Chronicle, with
D. C. Douglas and S. I. Tucker, Rutgers University
Press, New Brunswick, New Jersey, 1961.

William (of Malmesbury), Historia Novella, ca 1143.
There is a translation with notes by K. R. Potter,
Thom. Nelson and Sons, London, 1955.

Wylie, A., Chinese Researches, Shanghai, 1897. (No publisher's name given.)

PART TWO: THE ACCELERATIONS

CHAPTER VII

INTRODUCTION TO PART II

Part I of this study presented records and summaries of various types of ancient astronomical observations. Part II contains the quantitative analysis of the observations and the resulting estimates of the secular accelerations $\dot{n}_M$ of the moon's orbital motion and $\dot{\omega}_e$ of the earth's spin.

It will be assumed that the standard ephemerides of the sun, moon, and earth are so nearly correct that each observed quantity is a linear function, rather than a more complicated function, of the accelerations $\dot{n}_M$ and $\dot{\omega}_e$. Thus the ith observation provides a measurement, say Z_i, of some linear combination of $\dot{n}_M$ and $\dot{\omega}_e$. This measurement is subject to an error with a standard deviation σ_{Z_i}, say. Hence the ith observation provides an equation of condition that can be written in the form

$$A_i(\dot{n}_M + 22.44) + B_i(10^9 \, \dot{\omega}_e/\omega_e) = Z_i \pm \sigma_{Z_i}. \quad \text{(VII.1)}$$

In Part II, the units in which $\dot{n}_M$ is expressed will always be seconds of angle per century per century, and the units in which $10^9(\dot{\omega}_e/\omega_e)$ is expressed will always be reciprocal centuries. The quantity multiplied by A_i in Eq. (VII.1) is the difference between the acceleration $\dot{n}_M$ and the standard value $-22''.44/cy^2$.

The first goal of the quantitative analysis is to find values of A_i, B_i, Z_i, and σ_{Z_i} for each observation. After that goal is met, the values of $\dot{n}_M$ and $\dot{\omega}_e$ that best satisfy all the equations of condition will be found.

The data were organized in Part I partly by source and partly by type of measurement. In Part II they will be analyzed by type. The categories of data that will be used in Part II are: (a) lunar conjunctions and occultations, (b) planetary conjunctions and occultations, (c) magnitudes of partial lunar eclipses, (d) times of lunar eclipses,

(e) magnitudes of partial solar eclipses, (f) times of solar eclipses, and (g) positions of points near the paths of large solar eclipses.

These categories will be taken up in the order listed above; this is also approximately the order in which the analyses of the various categories were completed. In the inference of the accelerations the results from Chapter II, Part I, will be picked up.

The data will be arranged chronologically within each category. Each category will be broken into two parts, one for observations made before 500 Jan 0 and one for observations made since. This division will assist us in judging whether there has been a significant change in the accelerations within historic times. Within each category of data, the division largely, but not entirely, preserves arrangement by provenance. More time divisions will be used for well-populated categories.

Within several categories, the values of A_i and B_i are substantially the same for all observations. When this happens the equations of condition for the individual measurements can be combined into a smaller number of equations. In some cases the data from a category will be reduced to two equations, one for observations before and one for observations after 500 Jan 0. In other cases a single equation of condition will be found for all measurements within a category from a single provenance.

The B_i are zero only for the measurements of lunar eclipse magnitudes. For all other measurements we can divide Eq. (VII. 1) by B_i and find that $-(A_i/B_i) = d(10^9 \dot{\omega}_e/\omega_e)/d\dot{n}_M$. Except for measurements involving solar eclipses, the necessary derivatives can be evaluated with enough accuracy from simple principles. For solar eclipse measurements, the derivatives needed will be evaluated numerically, using eclipse calculations based upon both the standard value of $\dot{n}_M$ and upon a perturbed value.

I have written computer programs for calculating the circumstances of solar and lunar eclipses, using the

186

basic methods described in Chapter 9 of the Explanatory Supplement [1961]†. The methods used in these programs differed from those in ES in the following details: (a) instead of calculating time derivatives of certain quantities by numerical differentiation, I calculated them by approximate analytic methods; (b) I used only about the largest third of the perturbation terms given by Eckert et al. [1954] for the lunar ephemeris and by Newcomb [1895a] for the solar ephemeris‡; (c) in calculating local circumstances of solar eclipses, I used the "alternative method" described on page 244 of ES; (d) I made a provision whereby I could use an arbitrary acceleration of the moon in place of the standard value $-22''.44/\text{cy}^2$.

The staff of the Naval Observatory calculated the general circumstances of the solar and lunar eclipses, using their standard eclipse programs. Their results allowed me to test my programs thoroughly. The most delicate test is provided by the paths of solar eclipses. I found no path differences greater than about 45 km on the surface of the earth, and most differences were less than 10 km. I am grateful to the Observatory staff for these calculations, which were of great value in the preparation of my programs.

It is fairly common for the path of a solar eclipse to cut a given parallel of latitude at two longitudes. When this happens the value of Z_i in Eq. (VII. 1) is double-valued. When I felt that a choice between the values could be made

---

† References used in Part II are listed at the end of Part II even if they already appear in the list at the end of Part I. In the rest of Part II, this reference will be identified as ES for brevity.

‡ There are a number of typographical errors in the values of the perturbation coefficients given by Newcomb. The Naval Observatory has prepared a list of errata, which it intends to publish. I am indebted to Dr. R. L. Duncombe for a copy of the errata.

safely, I used one solution and ignored the other. When I felt that a choice could not be made, I did not use the record in inferring the accelerations.

In most cases I made the choice on the basis of the value of the parameter Y defined in Eq. (XIV.13), whose time history is well defined by Figure XIV. 2. In all cases in which the possible values of Y differed by 10 or more, one value was close to that given by Figure XIV. 2, and I adopted that value. When the values of Y were closer than 10, I put the record into the "ambiguous" category discussed in Chapter XIII. An ambiguous observation is not an observation of low accuracy.

In a few cases a record did not furnish a well-determined value of Y because the derivative dy/dx (see Section XIV. 5) was too far from the standard value 0. 622. These were cases in which the observer's distance from an eclipse path could not be changed appreciably by moving him in longitude. On the contrary, the observer's distance depended in these cases mostly upon $\dot{n}_M$. Ambiguity in these cases showed itself through a need for an extremely large value of $\dot{n}_M$ even though the observer was fairly close to the path to begin with. I put these cases into the ambiguous category also.

The values of (A_i / B_i) derived from elementary theoretical considerations, such as for the observations of lunar eclipse magnitudes, were tested by a few sample calculations with two values of $\dot{n}_M$.

CHAPTER VIII

ANALYSIS OF THE CONJUNCTIONS AND OCCULTATIONS

1. COMPUTATION OF STELLAR COORDINATES

Present positions and proper motions of the stars involved in the conjunctions and occultations were listed in Table VI. 1 with respect to the mean equator and equinox of 1950. 0. The first task in analyzing the conjunctions and occultations is to find the coordinates at the epochs needed.

The Islamic conjunctions were judged with respect to motion in longitude rather than in right ascension, so stellar ecliptic coordinates are needed in their analysis. The Ptolemaic observations can be calculated in either equatorial or ecliptic coordinates. Since the motion of the moon can be calculated more easily in ecliptic coordinates than in equatorial ones, the simplest coordinate system to use in the analysis of all the conjunctions and occultations is the mean ecliptic coordinate system of date.

The positions of the stars at the ancient epoch but in the fixed coordinate system that coincides with the mean equatorial system of 1950. 0 were first calculated from the values in Table VI. 1. In this calculation, the mean motions were taken constant. The coordinates were then transformed to the fixed system that coincides with the mean ecliptic system of 1950. 0 by rotation about the equinox through the angle 23°. 4458. This is the value of the obliquity calculated for the epoch 1950. 0 from the mean elements listed on page 98 of ES [1961][†]. The problem is now to transform ecliptic coordinates from the system for one epoch to the system for another.

[†] The mean elements on page 98 of ES were originally derived by Newcomb [1895a].

Formulas[†] for doing this are given at the bottom of page 38 of $\underline{ES}$. $\underline{ES}$ states near the middle of page 38 that the formulas may be used provided the time interval is not too long, but does not give an indication of how rapidly the accuracy degrades with time interval. In order to find how accurate the formulas are when applied over 22 centuries, which is about the longest interval needed in this work, I numerically integrated the exact differential equations of the transformation from the epoch 1950 Jan 0.5 backward for 22 Julian centuries. Care was taken to change the basic rates of motion from the amounts per tropical year, as they are given in $\underline{ES}$, to the amounts per Julian year. The precession during the interval from the epoch 1950 Jan 0.5 to the epoch 1950.0 = 1950 Jan $0^d.923$ was ignored.

Comparison of the rotation matrix obtained by the numerical integration and that which leads to the approximate formulas in $\underline{ES}$ shows that the error in using the approximate formulas does not exceed about 3" if the latitude does not exceed about 5°. This accuracy for this limitation in latitude is adequate for the purposes of the present study.

If θ_0, λ_0 denote the mean latitude and longitude of a star and if λ_S denotes the longitude of the sun, the aberrations in stellar latitude and longitude are approximately $- k_A \sin \theta_0 \sin (\lambda_S - \lambda_0)$ and $- k_A \sec \theta_0 \cos (\lambda_S - \lambda_0)$, respectively[‡]. k_A is the constant of aberration, given as 20".47 on page 48 of $\underline{ES}$. The quantities given when added to the mean position θ_0, λ_0 yield the apparent position Θ_0, Λ_0.

[†] It should be noted that the definitions of the quantities $\underline{c}$ and $\underline{c}'$ used in these formulas were interchanged accidentally where they were given near the center of page 38 of $\underline{ES}$.

[‡] I do not know the source of these approximations. I found them in Fotheringham [1915], verified them, and used them. Fotheringham gives no source for them, so they were probably once considered standard. Their derivation involves assuming uniform motion for the earth, the neglect of terms in $k_A{}^2$, and the approximation $\cos \theta_0 = 1$.

2. LUNAR CONJUNCTIONS AND OCCULTATIONS

The data concerning the lunar conjunctions and occultations preserved by <u>Ptolemy</u> [ca 152, VII. 3] were presented in Section VI. 2 and Table VI. 2. No other ancient observations of lunar conjunctions and occultations have been found.

The observations give the longitude of the moon at measured values of solar time. Their analysis poses a problem similar to that encountered in using the mean longitude of the moon in Section II. 4. The analysis can be performed using ephemeris time, but it can be done more easily using mean solar time. The immediate result of the latter choice is to give the quantity called $2L/\tau^2$ in Chapter I, which is the acceleration of the moon with respect to mean solar time. The accelerations $\dot{n}_M$ and $\dot{\omega}_e$ with respect to ephemeris time are related to $2L/\tau^2$ by Eqs. (I. 1). Since the ephemerides to be used in this section are based upon the standard value $\dot{n}_M = -22''. 44/cy^2$, $\dot{n}_M$ in Eq. (I. 1) should be replaced by $\dot{n}_M + 22''. 44/cy^2$.

The first step in the analysis of the data in Table VI. 2 was to find the geometric longitude and latitude of the moon from the tables in <u>Nav. Obs.</u> [1953], using the values of mean solar time from Table VI. 2 for the independent variable. The independent variable in the tables is usually considered to be ephemeris time. However, if there were no accelerations other than the standard set used in this work, mean solar time and ephemeris time would be identical, and we can use values of mean solar time as if they were ephemeris time for the purpose of this section.

The geometric longitude and latitude of the moon must be corrected for parallax before they can be used in connection with Table VI. 2. The distance to the moon, which is not given in <u>Nav. Obs.</u> [1953], is needed in calculating the parallax. Since the parallax is a fairly small correction, the distance is not needed accurately and simplified methods can be used to find it. I calculated the values of the distance by accepting the values of the horizontal equatorial parallax tabulated by Fotheringham [1915]. I then calculated the actual parallax for each observation, using distance found in this way and the geometric longitudes and latitude already found.

The effect of aberration upon the apparent position of the moon was ignored.

The positions of the cusps of the lunar crescent are needed for some observations. I accepted Fotheringham's calculations of the apparent position of the cusps with respect to the apparent position of the center of the moon. Since one would make an almost negligible error by assuming that the cusps have the same longitude as the center, I have not considered it necessary to verify the relative cusp positions.

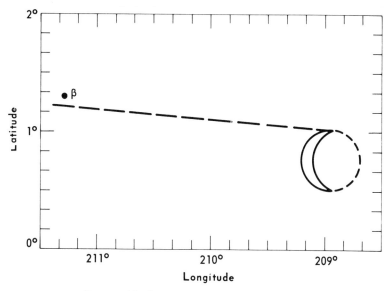

Figure VIII. 1. Calculated positions of the moon and of 8 β¹ Sco on −294 Dec 21.

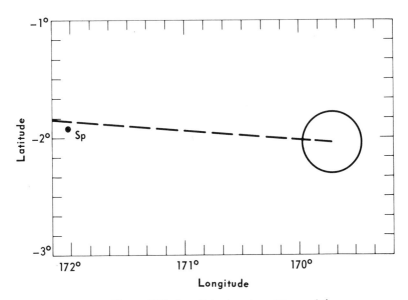

Figure VIII. 2.   Calculated positions of the
moon and of Spica on −293 Mar 9.

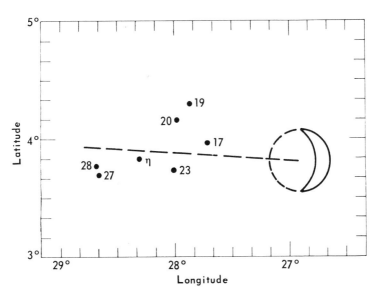

Figure VIII. 3.   Calculated positions of the
moon and the Pleiades on −282 Jan 29.

193

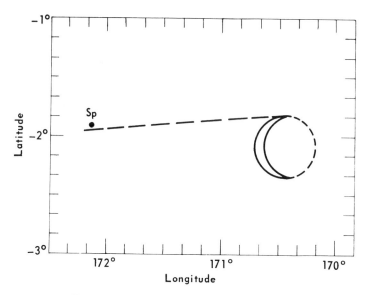

Figure VIII. 4.   Calculated positions of the
moon and Spica on −282 Nov 9.

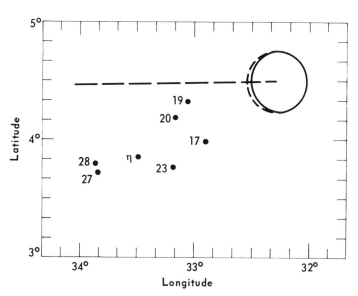

Figure VIII. 5.   Calculated positions of the
moon and of the Pleiades on 92 Nov 29.

194

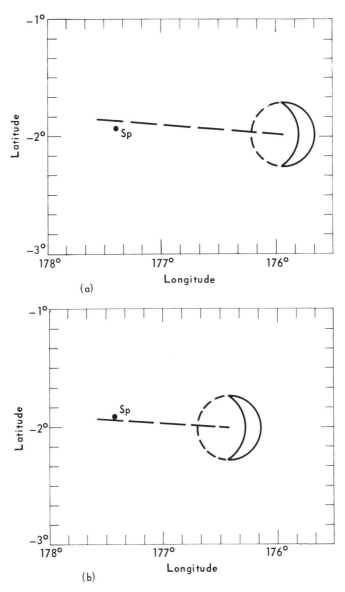

(a)

(b)

Figure VIII.6.   (a) Calculated positions of the moon and of
Spica on 98 Jan 11 at a time when Spica was observed to be
occulted. (b) Calculated positions of the moon and of Spica
about an hour later.

195

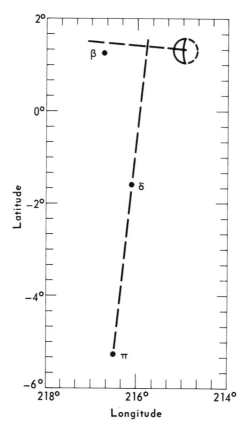

Figure VIII. 7.   Calculated configuration of the
moon and of part of the Scorpion on 98 Jan 14.

The apparent positions of the moon calculated in the
way just described are plotted for each observation in Figures VIII. 1 through VIII. 7; the independent observations connected with the sixth record are given independently in Figures VIII. 6a and VIII. 6b.   The phases of the moon, like the
other quantities that are not central to the use of the data,
were accepted from Fotheringham [1915].   The apparent
stellar positions found in Section VIII. 1 are also plotted in
Figures VIII. 1 through VIII. 7.   The stars are identified in
the figures by using only small but characteristic parts of
the symbols given them in Table VI. 1. $8\beta^1$ Sco, for example, is marked β. 67α Vir is identified by Sp for Spica.

The stellar positions plotted agree closely with those given by Fotheringham, except that the latitudes of the three stars involved in the observation of 98 Jan 14 found here are all smaller by about 3′. The interpretation of the observations is not appreciably affected by this small discrepancy. The lunar positions found here, whether geometric or apparent, show considerable discrepancy with those found by Fotheringham. Most of the discrepancy comes from the difference between standard values used for the lunar acceleration. However, there is an additional discrepancy of the order of 0°.1 which can have either sign. I do not know the origin of this discrepancy, but it probably comes from different methods of calculating lunar perturbations.

TABLE VIII. 1

ANALYSIS OF LUNAR CONJUNCTIONS AND OCCULTATIONS
PRESERVED BY PTOLEMY

Date	Δ (deg)	A priori Error Estimates				$2L/\tau^2$ " $/cy^2$	A priori Stan. Dev. of $2L/\tau^2$ " $/cy^2$
		Record (deg)	Angle Estimate (deg)	Time Rounding (hr)	Clock Error (hr)		
- 294 Dec 21	2.40	0	0.05	0.29	0.09	40.2	2.6
- 293 Mar 9	2.00	0	0.05	0.29	0.06	32.8	2.5
- 282 Jan 29	1.65	0.05	0.05	0.29	0.09	24.1	2.7
- 282 Nov 9	1.72	0.03	0.05	0.29	0.07	23.8	2.6
+ 92 Nov 29	1.42	0.05	0.05	0.29	0.06	34.6	4.0
98 Jan 11a	1.48	0.17	0.05	0.29	0.06	32.2	5.3
98 Jan 11b	1.43	0.07	0.05	0.29	0.03	31.1	4.0
98 Jan 14	1.27	0	0.05	0.29	0.03	29.8	3.7

The direction in which the calculated position of the moon would be changed by changing the secular acceleration is shown by the dashed lines in the figures. This direction is not quite the same as the apparent direction of motion of the moon. The latter depends upon the change in parallax, which involves the rotation of the earth, as well as upon the orbital motion of the moon.

The column headed Δ in Table VIII. 1 gives the changes in the longitude of the moon needed to make the calculated configurations (shown in Figures VIII. 1 through VIII. 7) agree with the observations. Several observations

197

allow a range of Δ. For example, in the observation 98 Jan 11a, Spica is occulted if Δ lies between 1°. 20 and 1°. 77; allowance is made for the effect of a "bright" occultation, as mentioned in Section VI. 2. If a range of Δ is allowed, the central value of Δ is listed under Δ. The range, divided by 2/3, is listed in the column "Record (deg)" in Table VIII. 1. This amount is proportional to the contribution of the uncertainty in lunar position given by the record to the a priori estimate of the error associated with the observation.

Three other estimated error contributions are listed in Table VIII. 1. The constant value 0. 05 listed under "Angle Estimate (deg)" is an estimate of the error that an observer would make in estimating a celestial position, or in estimating whether two objects had the same position. This value is based to a considerable extent upon the results to be found in the next section. The constant value 0. 29 listed under "Time Rounding (hr)" is the standard deviation of the error in rounding time measurements to the nearest half-hour. The value listed under "Clock Error (hr)" is 0. 05 times the interval from the recorded time to sunrise or sunset, whichever is nearer.

The acceleration $2L/\tau^2$ that corresponds to the value of Δ is given in the seventh column of Table VIII. 1. It is not simply 2Δ divided by the square of the age of the observation (before the epoch 1900 Jan 0. 5) and converted to the units given. The reason is that a change in the mean longitude of the moon produces a different change in the position of the moon according to where the moon is in its orbit. The moon is displaced more near perigee than near apogee. For the derivative $d\Delta/dL$, I used simply

$$d\Delta/dL = 1 + 2e_M \cos M,$$

in which e_M is the eccentricity of the lunar orbit and M is the mean anomaly. This ignores many small perturbations but is correct to the order of 0. 01; the average error in it is zero.

The contributions of the four errors already discussed when converted to their equivalents in $''/cy^2$ were

198

found next. The two angular errors listed affect Δ directly and were converted as they are. The time errors were multiplied by the mean motion of the moon; this gives their effect upon the mean longitude. The effect upon the acceleration was obtained upon further multiplication by $2/\tau^2$ and conversion to the units given. The number listed as "A priori Stan. Dev. of $2L/\tau^2$, $''/cy^2$" is the square root of the sum of the squares of the four individual contributions.

A weight equal to the "interpretation factor" in Table VI. 2 divided by the square of the a priori standard deviation was assigned to each value of $2L/\tau^2$ in Table VIII. 1. The weighted mean of $2L/\tau^2$ is

$$\overline{2L/\tau^2} = 30''.6/cy^2 .$$

The deviations of the individual values of $2L/\tau^2$ from the mean are of the same general size as the a priori estimates, but there is no noticeable correlation. In fact, the observations made in the -3rd century have the lowest a priori estimates but the largest deviations. This suggests that the disadvantage of a smaller value of τ^2 for the more recent observations is at least balanced and may be outweighed by the advantage of better techniques. The a priori estimates also indicate that the largest error sources in the first century observations are probably the range of Δ allowed by the observation and the time rounding. These should not be correlated between the two observations on 98 Jan 11, and we are probably justified in treating these as independent observations, contrary to the expectation implied in Section VI. 2.

The deviations suggest that equal weights are as justifiable as the weights based upon the a priori estimates. The estimate found by assigning equal weights is

$$2L/\tau^2 = 31''.1 \pm 1''.9 \text{ cy}^{-2} . \qquad (VIII. 1)$$

The standard deviation given in Eq. (VIII. 1) is that found in the usual statistical way from the scatter of the values about the mean. The mean in Eq. (VIII. 1) does not differ significantly from the weighted mean already found and is at least as well justified. Therefore, I propose the value in Eq. (VIII. 1) as the best estimate to be found from the lunar conjunctions and occultations preserved by Ptolemy.

Fotheringham [1915] found $4''.5 \pm 0''.7 \ cy^{-2}$ for L/τ^2, which he calculated with respect to the condition of no secular accelerations. The value found here is referred to the value $\dot{n}_M = -22''.44 \ cy^{-2}$. Fotheringham's value is equivalent to $2L/\tau^2 = 31''.4 \pm 1''.4$ in the conventions used here. He quoted probable errors rather than standard deviations. It is encouraging that there is little difference between Fotheringham's estimate and the estimate found here, in spite of considerable differences with regard to some of the individual observations.

3. PLANETARY CONJUNCTIONS AND OCCULTATIONS

The data concerning the Venus conjunctions and occultations preserved by Ebn Iounis [1008] were presented in Section VI. 3 and Table VI. 3. Since the analysis of these records involves the earth and planets only, and not the moon, the records furnish an estimate of $\dot{\omega}_e$.

The central task in analyzing these data is to calculate the values of ephemeris time (ET) at which the observed planetary positions occurred. In order to do this, the positions of the planets involved in each observation were calculated for a value of ephemeris time equal to the stated value of Greenwich mean solar time (GMST)[†] when the observation was made. The angular velocities of the planets were also calculated for the same time. A value of $\Delta t = GMST - ET$ can then be calculated for each observation from these quantities. The positions and velocities were calculated with respect to the mean ecliptic coordinate system of date.

[†] The reason for using this symbol, which is often used for sidereal time, is given in a footnote in Chapter I.

The method of calculating the stellar coordinates has been described in Section VIII. 1. The angular velocities of the stars with respect to the mean ecliptic coordinate system of date are ignored in the final step of calculating Δt.

The heliocentric coordinates of Mercury, Venus, and Mars were calculated using the expressions[†] of Newcomb [1895b, 1895c, 1898] for the mean elements of these planets, neglecting the perturbations about the positions calculated from the mean elements. Newcomb's tables for these planets, given in the references just cited, consist of tables of the various perturbations about the position given by the mean elements. In order to make the tabulated quantities positive, Newcomb added a constant to the quantity in each table; the sum of all the constants could then be subtracted out as the last step in using the tables. The sums of all the constants are $40''$ for Mercury, $35''$ for Venus, and $127''$ for Mars.

These numbers are comparable with the amounts that the planets move in the time interval Δt. Nevertheless, it was considered safe to neglect the perturbations. The numbers listed above are bounds to the perturbations. They are attained by the actual perturbations only if the oscillatory functions tabulated in approximately ten different tables attain their extrema simultaneously. This is unlikely, hence the root-mean-square values of the perturbations are probably much smaller than the values listed. Further, the mean values of the perturbations are zero, so that neglecting them should not cause a bias.

The heliocentric longitudes of Jupiter and Saturn were calculated, using the tables of Hill [1895a, 1895b]. The calculations were performed to a precision of a few seconds of arc. For simplicity the latitudes and radii vectores, which are needed only to low accuracy, were not taken from the tables. I calculated them by using

---

[†]These expressions are reproduced on page 113 of ES.

approximate expressions for the eccentricity e, the longitude of perihelion $\widetilde{\omega}$, the inclination $\underline{i}$, and the longitude of the node Ω. It is not easy to find expressions for the secular variations of the mean elements of the outer planets; this was the reason for using Hill's tables rather than mean elements for Jupiter and Saturn. The expressions I used are

$$\underline{e} = 0.048\ 337 + 0.000\ 165\ 1T,$$

$$\widetilde{\omega} = 12°.7123 + 1°.60981T,$$

$$\underline{i} = 1°.3088 - 0°.00569T,$$

$$\Omega = 99°.4381 + 1°.01057T$$

(VIII.2)

for Jupiter and

$$\underline{e} = 0.055\ 889 - 0.000\ 344\ 5T,$$

$$\widetilde{\omega} = 91°.0898 + 1°.95771T,$$

$$\underline{i} = 2°.4923 - 0°.00389T,$$

$$\Omega = 112°.7839 + 0°.87262T$$

(VIII.3)

for Saturn. I found these by combining the elements given for the epoch 1850 Jan 0, Greenwich Mean Noon, by Hill with those given for the epoch 1958 Jan 0, GMN on page xvii of the American Ephemeris and Nautical Almanac [1958]. The accuracy of Eqs. (VIII.2) and (VIII.3) over a span of a millenium is certainly not high but it should be adequate for the use to which they are put.

The geocentric coordinates of the sun are needed in order to transform the coordinates of the planets from heliocentric to geocentric. The latitude of the sun was taken to be zero. The longitude of the sun was taken from Nav. Obs. [1953]. The eccentricity and the longitude of perigee were taken from the expressions on page 98 of ES.

202

From these quantities the true anomaly and thence the earth-sun distance were calculated.

The heliocentric angular velocities of Jupiter and Saturn were found by taking differences in Hill's tables. To find the angular velocities of the minor planets, including the earth, I assumed that Kepler's Second Law[†] holds exactly and calculated the angular velocities from the radii. To find the geocentric angular velocity $\dot{\Lambda}_p$ of another planet from the heliocentric angular velocities $\dot{\lambda}_p$ and $\dot{\lambda}_E$, I used the expression[‡]

$$R_p^2 \dot{\Lambda}_p = \dot{\lambda}_p r_p^2 + \dot{\lambda}_E r_E^2 - (\dot{\lambda}_p + \dot{\lambda}_E)(\underline{r}_p \cdot \underline{r}_E). \quad \text{(VIII. 4)}$$

In this, R_p is the geocentric radius of the planet, $\underline{r}_p$ is its heliocentric radius vector, and $\underline{r}_E$ is the heliocentric radius vector of the earth.

Parallax was neglected in finding the apparent planetary positions from the geocentric positions. Aberration in the case of a planet was allowed for by subtracting $498^S.38 R_p \dot{\Lambda}_p$ from the geocentric longitude to yield the apparent longitude; in using this, the distance R_p is to be measured in astronomical units. The aberration in latitude was neglected. The method of finding stellar aberration has already been described in Section VIII. 1.

[†] That is, the law that says that the radius vector of a planet sweeps equal areas in equal times.

[‡] It is so simple to derive Eq. (VIII. 4) that I have not attempted to find a source for it. Its derivation requires neglecting $\sin^2 i$ of a planet compared with unity. Since $\dot{\Lambda}_p$ is hardly needed with an accuracy of better than 1 part in 10, this neglect is legitimate.

TABLE VIII.2

CALCULATED LATITUDE DIFFERENCES WHEN THE OBSERVED
LATITUDE DIFFERENCE WAS ZERO

Tabular Date	Calculated Lat. of Venus Minus Lat. of Second Body (deg)
864 Feb 13	-0.10
864 Oct 10	-0.03
885 Sep 9	-0.04
996 Aug 8	0.08
997 May 24	-0.05
1002 Apr 18	-0.03

It is convenient to dispose of a minor point concerning the Islamic conjunctions and occultations before going to the main results. Six records assert that the difference in latitude between Venus and the other body was zero, by saying that they seemed to touch, that there was a perfect occultation, or the like. The calculated differences in latitude for these six records are listed in Table VIII.2. The mean value of the difference is -0°.028. The standard deviation of the difference is 0°.061; since there are only six values in the sample, the mean is not significantly different from zero.

This result suggests that 0°.06 is about the limit of the astronomers for resolving differences in position or for measuring small differences in position. It is fairly close to the estimate of 5 ' made in Section VI.3 on the basis of the precision apparently used by the astronomers in reporting positions. It is also the basis for the error contribution of 0°.05 used in the fourth column of Table VIII.1. Since the exact value is neither important nor well established, the value was rounded from 0°.06 to 0°.05 in Table VIII.1.

We now turn to the main point. The dates of the observations are listed in the first column of Table VIII.3.

TABLE VIII. 3

ACCELERATION OF THE EARTH'S SPIN AS INFERRED FROM THE
ISLAMIC RECORDS OF CONJUNCTIONS AND OCCULTATIONS

| Date | GMST from Noon (hr) | Long. of Venus Relative to Second Body (deg) | | Ang. Velocity of Venus Relative to Second Body (deg/day) | Δt^a (hr) | 10^9 x $(\dot\omega_e/\omega_e)$, cy^{-1} |
		Observed	Calculated with $\dot\omega_e = 0$			
858 Aug 28	- 9.4	0.4W	- 0.44	1.11	- 0.9	- 19
864 Feb 13	2.7	0	0.07	0.49	3.3	71
864 Oct 22^b	- 9.0	0	- 0.02	0.53	- 0.9	- 18
885 Sep 10^b	-10.8	0	- 0.02	1.19	- 0.4	- 8
901 Oct 4	- 9.6	0	- 9.63	1.02	-226.2	--
918 Dec 24	- 9.4	4.48E	4.60	0.44	6.4	152
987 Jun 17	5.9	0	- 0.25	0.87	- 6.8	-187
988 Jan 20	- 7.7	0	+ 0.05	1.12	1.1	30
990 Jun 23^b	5.8	0	- 0.36	1.15	- 7.5	-207
991 Dec 22	3.9	?	- 0.27	1.13	- 5.8	c
992 Sep 17^b	- 9.3	0.33E	0.40	1.14	1.3	37
995 Jan 3	3.1	1W	0.04	-0.48	- 51.8	--
995 Jun 11	4.9	0	- 0.06	0.72	- 1.9	- 52
995 Jun 18	5.6	0	0.32	0.83	9.2	257
996 Aug 15^b	5.2	0	- 0.07	1.02	- 1.6	- 44
997 May 24	- 9.7	0	0.03	0.76	1.1	30
998 Jun 4	5.9	0.25E	-16.55	0.58	-700.8	--
998 Jun 23	6.0	0	0.04	1.14	0.8	22
999 Apr 10	- 9.6	1E^b	0.96	0.42	- 2.0	- 57
1000 May 19	5.9	0	- 0.04	-1.58	0.5	15
1001 May 28^b	9.9	0	0.79	1.42	13.4	--
1002 Apr 18	-10.1	0.2W	- 0.29	0.89	- 2.5	- 70
1002 Jul 15	- 9.7	0.33E	0.62	1.09	6.3	177
1003 Jan 7	10.0	0	0.02	0.46	1.1	31
1003 Feb 19	-11.8	0	0.13	1.01	3.0	85
1003 Jun 18	11.9	0	0.20	0.80	6.1	172

^{a}Solar time - ephemeris time.

^{b}Differs by an integer from the listing in Table VI. 3. See main text.

^{c}Not used. See Section VI. 3.

The times of the observations, converted[†] to GMST with
respect to noon on the date tabulated, are given in the
second column; a negative value means that the observation
was before noon. The observed longitudes of Venus relative
to the second body are listed in the third column.

[†]Here and elsewhere, when it was necessary to find the
time of sunrise or sunset in converting times, I used the
table of sunrise and sunset in American Ephemeris and
Nautical Almanac [1967, pp. 384-391]. I assumed that the
table would be the same every year if the date were

The entries marked with a superscript b in the first or third columns of Table VIII. 3 differ by an integer from the corresponding entries in Table VI. 3. The dates were changed because calculation with the date listed in Table VI. 3 gave values of Δt that were almost exactly equal to an integral number of days. The likelihood that this would happen in several cases was pointed out in Section VI. 3. In some cases the numbers used in Tables VI. 3 and VIII. 3 are different but the epochs specified are the same; these cases are not marked. The observation of 1003 Feb 18/19 furnishes an example. In the third column for the observation of 999 Apr 10, 1° W in Table VI. 3 has been changed to 1°E in Table VIII. 3. This change was made because 1°W leads to an unreasonable result while 1°E leads to a reasonable one[†].

---

expressed in the Gregorian calendar. The error made because of this assumption is small compared with the errors of observation. If the observations are spread uniformly through the time of year, the algebraic mean error made is zero.

[†]The record, which is quoted in Section VI. 3, says that Venus preceded Mars by about 1°. It was noted in Section VI. 3 that similar wording was used for the record of 995 Jan 3. It was also noted in the discussion of the latter record, but not repeated in the discussion of 999 Apr 10, that "to precede" usually meant "to be west of" in ancient records, but that the meaning was ambiguous in the context under discussion. The record of 995 Jan 3, as it stands in Tables VI. 3 and VIII. 3, leads to $\Delta t = -51^h. 8$, which is unreasonable. Changing the date by two days would give a reasonable value of Δt. However, the need to change the date, when combined with the ambiguity in the sign of the longitude difference, would lead to low confidence in the result, and I have simply omitted the observation of 995 Jan 3 from the final inference.

The fourth and fifth columns give the longitude and longitude rate (angular velocity) of Venus relative to the other body, calculated for a value of ET equal to the value of GMST specified by the first two columns. A positive value in the longitude column means that the calculation placed Venus east of the other body. A positive value in the angular velocity column means that Venus is traveling eastward with respect to the other body. The values of Δt = GMST - ET can be calculated from the fourth and fifth columns.

A value of GMST given by the first two columns of Table VIII. 3 plays two roles. First and obviously, it is the observed value of solar time when the specified planetary configuration occurred. Second, it has been used as a first approximation to the value of ET at the time of the specified configuration. It will help in the discussion to use different symbols for the two roles. For the present let $(ET)_o$ denote an approximate value of ET, and forget how $(ET)_o$ has been chosen. Let Λ_{VO} and $\dot{\Lambda}_{VO}$ be the apparent longitude and longitude rate of Venus calculated for the epoch $(ET)_o$. The longitude Λ_V of Venus at a neighboring epoch (ET) is

$$\Lambda_V = \Lambda_{VO} + [(ET) - (ET)_o] \, \dot{\Lambda}_{VO} \, .$$

The longitude of the second body is given by a similar expression, with the quantities appropriate to Venus being replaced by the corresponding quantities for the second body.

The quantity Δt that we use in estimating $\dot{\omega}_e$ is defined by Δt = (GMST) - (ET). If we now remember that $(ET)_o$ is chosen to equal (GMST), the expression for Λ_V becomes $\Lambda_{VO} - \dot{\Lambda}_{VO} \Delta t$, and the expression for Λ_2, the apparent longitude of the second body, becomes $\Lambda_{20} - \dot{\Lambda}_{20} \Delta t$. Suppose for simplicity that the observed longitude difference in Table VIII. 3 is zero. Then the desired value of Δt is the one that makes $\Lambda_V = \Lambda_2$, or

$$\Delta t = (\Lambda_{VO} - \Lambda_{20})/(\dot{\Lambda}_{VO} - \dot{\Lambda}_{20}). \qquad \text{(VIII. 5)}$$

The modification needed if $\Lambda_V \neq \Lambda_2$ is obvious. The values of Δt calculated from Eq. (VIII. 5) and converted to hours are listed in the sixth column of Table VIII. 3.

The values of Δt for the observations of 901 Oct 4 and 1001 May 28 are close to $-9\frac{1}{2}$ and to $\frac{1}{2}$ days, respectively. Since we cannot resolve an error of 1 day on the basis of the calculations, these records will be dropped. The date for the second of these had already been changed from 1001 Jun 2 because calculation with the latter date indicated that Δt was close to an integral number of days. However, the relative angular velocity changed so rapidly that the original calculation of Δt was in error by nearly 12^h.

The record of 995 Jan 3 will be dropped for the reason discussed in a footnote above. The record of 991 Dec 22 is ambiguous and was assigned zero weight in Section VI. 3. It was calculated only to see if the meaning of the passage could be guessed. The best guess on the basis of Table VIII. 3 is that the observer meant to report an actual conjunction, but this is not the only possibility. Further study of this record does seem warranted here.

The record of 998 Jun 4 is apparently in error by almost a month. It should perhaps be recalculated for 998 Jul 4, but it hardly seems worth the trouble. This record will be discarded.

The last column in Table VIII. 3 gives the values of $\dot{\omega}_e$ calculated from

$$\dot{\omega}_e = 2\Delta t/T^2. \qquad \text{(VIII. 6)}$$

This is Eq. (II. 5) of Part I repeated in different notation. It should be remembered that T is time in Julian centuries of 36 525 ephemeris days from the fundamental epoch 1900 Jan 0.5 ET = JED 2 415 020.0. Values of $\dot{\omega}_e$ are not given for the records that have been discarded.

208

The estimate[†] of $10^9(\dot{\omega}_e/\omega_e)$ formed from the last column in Table VIII. 3 is

$$10^9(\dot{\omega}_e/\omega_e) = 20 \pm 24 \text{ cy}^{-1}. \qquad \text{(VIII. 7)}$$

The error estimate is the standard deviation and is based upon the scatter of the values in Table VIII. 3.

The error estimate in Eq. (VIII. 7) is disappointing. In Section VI. 3, I estimated that the error should be that corresponding to about $0^h.4$ in Δt. For the average age of the observations, this would give an error of about 10 cy^{-1} in $10^9(\dot{\omega}_e/\omega_e)$.

A histogram of the values in Table VIII. 3 suggests that two different populations of values are involved. Schroader [1968] has confirmed this and has shown part of their origin. It is plausible that the longitude error is correlated with the latitude difference. Following Schroader, I have plotted the longitude error against the latitude difference in Figure VIII. 8. This plot suggests that the expected correlation does exist to some extent. However, the most important conclusion that Schroader found comes from studying the directions of the arrow-heads on the symbols. If the arrowhead points up, the observation was made in the morning; if it points down, Venus was an evening star.

Only one morning observation shows an error exceeding $0°.1$. Except for this, all large errors were made in the evening. Such a difference between morning

---

[†]Each value of $10^9(\dot{\omega}_e/\omega_e)$ in Table VIII. 3 is assigned equal weight in reaching the estimate in Eq. (VIII. 7). There are reasons, some of which were discussed in Part I, for assigning slightly different weights to the observations. The estimate formed is hardly affected by using variable rather than uniform weights, so uniform weights are used for simplicity, in spite of statements made earlier.

and evening results could have many sources. If the instruments were sensitive to thermal gradients, they would be more distorted in the evening after exposure to sun all day. The observer on the observatory's morning shift could have been more skilled than the one on the evening shift. There are other possibilities. Without more study than is warranted here, we cannot find a convincing explanation of the difference.

The discovered difference affords no sound basis for discarding values from Table VIII.3. Therefore Eq. (VIII.7) will be taken as the best estimate that can be found from the Venus observations preserved by Ebn Iounis. The mean in Eq. (VIII.7) differs by about 40 cy^{-1}, or about $1\frac{1}{2}$ standard deviations, from the values given by other measurements. The probability of such an error happening by chance is about 0.134.

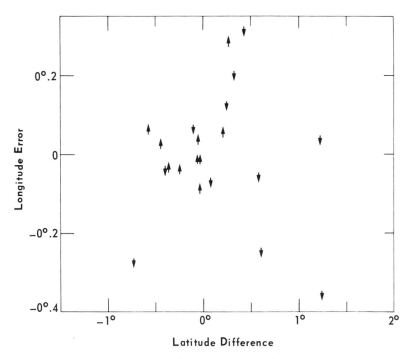

Figure VIII.8.   Longitude error in reported conjunctions of Venus as a function of separation in latitude. The arrow head points up for morning observations and down for evening observations.

CHAPTER IX

ANALYSIS OF THE MAGNITUDES OF PARTIAL
LUNAR ECLIPSES

1. GENERAL COMMENTS ABOUT MAGNITUDES OF LUNAR ECLIPSES

In this chapter, the term "magnitude", when used without qualification, will mean what was called the "magnitude of the diameter" in Chapter V. There will be little further need to use the magnitude of the area.

The magnitude of a lunar eclipse was defined as the fraction of the diameter of the moon lying within the earth's umbra. The magnitude defined in this way is a function of time that exists only when the moon is partially eclipsed. In order to write a formula for it, let μ denote the magnitude, s_M the angle subtended by the semidiameter[†] of the moon, λ the angle subtended by the radius of the earth's shadow at the distance of the moon from the earth, and m the angular separation between the centers of the moon and the umbra. Then

$$\mu = (s_M + \lambda - m)/2s_M . \qquad (IX.1)$$

All the quantities in the right member of Eq. (IX.1) are continuous functions of time, hence μ as defined by Eq. (IX.1) is a continuous function of time. Further, it agrees with the earlier definition of magnitude when the moon is partially eclipsed. Thus Eq. (IX.1) furnishes a useful generalization of the earlier definition. μ as defined by Eq. (IX.1) has the following properties:

[†]"Semidiameter" is commonly used in place of "radius", when speaking of the lunar disk, in order to let "radius of the moon" denote briefly the more frequently used quantity "radius vector of the center of the moon from the center of the earth". Similar usage is adopted for the sun. s_M is usually called simply the "semidiameter of the moon".

$\mu \leq 0$ when the moon is not eclipsed;

$0 < \mu < 1$ when the moon is partially eclipsed;   (IX. 2)

$\mu \geq 1$ when the moon is totally eclipsed.

Since there is no way to observe the edge of the earth's shadow except when it intersects the moon's disk, measured values of μ are always between 0 and 1.

If an observer on earth can see the moon, the value of μ that he will measure does not depend upon where he is on the earth's surface, except possibly for small parallax effects that will be neglected here. Hence anciently measured values of μ furnish no information about $\dot{\omega}_e$. It is useful to estimate next how μ_m, the maximum value of μ attained during an eclipse, depends upon $\dot{n}_M$, since all the measured magnitudes given in Chapter V were measurements of $\mu_m{}^\dagger$.

In evaluating $\partial \mu_m / \partial \dot{n}_M$, it is convenient to evaluate first the derivative $\partial \mu_m / \partial \Omega$, the rate at which μ_m would change if the position Ω of the lunar node were changed without changing the mean longitude L_M of the moon. With fair accuracy, the middle time of an eclipse is unchanged by such a change in Ω.

The magnitude μ_m depends upon the distance from the axis of the shadow to the plane of the lunar orbit at the

---

$\dagger$ Fotheringham [1909; also 1920, p. 124] assumed, without explanation that I noticed, that lunar eclipse magnitudes give an estimate of $2L'/\tau^2$, the "acceleration of the sun", as this quantity was defined in Chapter I. We can conclude that no typographical error is involved since the assumption is stated in several places in words, symbols, and tabular entries. The combination of this assumption with Eqs. (I. 1) would lead to the paradoxical result that lunar eclipse magnitudes depend only upon the spin acceleration $\dot{\omega}_e$ of the earth and not at all upon the orbital acceleration $\dot{n}_M$ of the moon.

time of opposition. Since the time is held fixed in evaluating $\partial\mu_m/\partial\Omega$, the change in the (angular) distance from the shadow axis to the orbital plane because of a change $\delta\Omega$ in Ω is $\delta\Omega \sin i_M$; i_M is the inclination of the lunar orbit. Hence

$$|\partial\mu_m/\partial\Omega| = \sin i_M/2s_M . \qquad\qquad \text{(IX. 3)}$$

With the aid of a simple sketch, the reader can easily convince himself that the sign of $\partial\mu_m/\partial\Omega$ can be found from the following rule: Let S_S be the sign of the sum of the declinations of the sun and moon at the time of opposition, and let S_Ω be the "sign of the node"; S_Ω is + if the moon is near its ascending node and - if the moon is near its descending node. The sign $S(\partial\mu_m/\partial\Omega)$ of $\partial\mu_m/\partial\Omega$ is given by:

$$S(\partial\mu_m/\partial\Omega) \text{ is + if } S_S \text{ and } S_\Omega \text{ are alike;}$$
$$\qquad\qquad\qquad\qquad\qquad\qquad\qquad\qquad \text{(IX. 4)}$$
$$S(\partial\mu_m/\partial\Omega) \text{ is - if } S_S \text{ and } S_\Omega \text{ are different.}$$

$\sin i_M = 0.0898$ and the value[†] of s_M at mean distance is [ES, p. 109] $15' 32''.58 \approx 0°.259$. Hence

$$|\partial\mu_m/\partial\Omega| = 0.173 \text{ deg}^{-1} \text{ at mean distance.} \qquad \text{(IX. 5)}$$

Turning to the derivative $\partial\mu_m/\partial\dot{n}_M$, we observe that a change $\delta\dot{n}_M$ in $\dot{n}_M$ changes L_M by the amount $\frac{1}{2}\delta\dot{n}_M T^2$ for a given value of ephemeris time[‡]. On the average the time of opposition is changed by the amount $-\delta\dot{n}_M T^2 \div 2(n_M - n_S)$; the sign means that opposition occurs earlier if $\delta\dot{n}_M > 0$. The mean longitude L_S of the sun at opposition thus changes by $-\frac{1}{2}\delta\dot{n}_M T^2 \times [n_S/(n_M-n_S)]$. The effect upon μ_m is the same as if Ω had changed by an equal but opposite amount. Insertion of numerical values yields

---

[†] Strictly speaking, $\sin s_M$ should be used in place of s_M.

[‡] T means time from the epoch 1900 Jan 0.5 measured in Julian ephemeris centuries, as it did in Part I.

$$\partial \mu_m / \partial \dot{n}_M = 0.\,000112 T^2 (\partial \mu_m / \partial \Omega) \; \mathrm{cy}^2 / '' \qquad \text{(IX. 6)}$$

if $\delta \dot{n}_M$ is in $''/\mathrm{cy}^2$ and $\delta \Omega$ is in degrees.

It is frequently stated [Jeffreys, 1962, p. 237, for example] that the torques that produce $\dot{n}_M$ have no effect upon the position of the node and hence that Eq. (IX. 6) gives the complete derivative of μ_m with respect to $\dot{n}_M$. This matter needs a brief investigation. If $\delta n_M / n_M \neq 0$, the semimajor axis of the lunar orbit must change, and hence the precession rate Ω must change. Approximately, $\delta \dot{\Omega} / \dot{\Omega} = (7/3)(\delta n_M / n_M)$. Insertion of numerical values shows that $\delta \Omega \approx 0.\,01\,\delta L_M$; further, $|\partial \mu_m / \partial \Omega| \approx 12 |\partial \mu_m / \partial L_M|$. Thus the value of $\partial \mu_m / \partial \dot{n}_M$ must be changed by about 12 percent from the value in Eq. (IX. 6) if we wish the partial derivative to include the indirect effect upon Ω as well as the direct effect upon L_M. It will appear later that the error produced by neglecting the indirect effect upon Ω is negligible compared with the errors in the ancient observations of μ_m.

2. DEDUCTIONS FROM THE OBSERVED MAGNITUDES

The observed magnitudes and some relevant results of computations are summarized in Tables IX. 1 and IX. 2. Table IX. 1 deals with observations made before the year 500 and Table IX. 2 with observations since then.

In each table the first column gives the eclipse identification used in Part I, including the provenance. The next two columns give respectively the observed magnitude μ_m and the value of μ_m calculated using the standard acceleration $\dot{n}_M = -22''.\,44/\mathrm{cy}^2$. The fourth column gives the sign of $\partial \mu_m / \partial \Omega$, obtained with the aid of Eq. (IX. 4) and of the elements and circumstances of the eclipses. The numerical values of $\partial \mu_m / \partial \Omega$ and of $\partial \mu_m / \partial \dot{n}_M$ can be found so simply from Eqs. (IX. 5) and (IX. 6) that it seems unnecessary to tabulate them. The last two columns in the tables need more extensive discussion.

214

TABLE IX. 1

MAGNITUDES OF LUNAR ECLIPSES OBSERVED BEFORE 500

Identification		Magnitude		Sign of $\partial\mu_m/\partial\Omega$	x_2 $''/cy^2$	$\Delta\Omega$ (deg)
		Observed	Calc.			
- 719 Mar 8	BA	0.317^b	0.127	+	92^c	0.71^c
- 719 Sep 1	BA	0.542^b	0.506	+	- 4c	-0.03^c
- 620 Apr 22	BA	0.250	0.141	-	- 88	-0.63
- 522 Jul 16	BA	0.500	0.525	-	22	0.14
- 501 Nov 19	BA	0.250	0.186	-	- 57	-0.37
- 490 Apr 25	BA	0.250^b	0.084	-	$- 75^c$	-0.48^c
- 382 Dec 23	BAa	0.142	0.202	-	--	0.35
- 173 Apr 30	M	0.583	0.627	+	- 52	-0.25
- 140 Jan 27	M	0.250	0.268	+	- 23	-0.10
+ 125 Apr 5	M	0.167	0.144	-	- 39	-0.13
134 Oct 20	M	0.833	0.839	+	- 10	-0.03
136 Mar 6	M	0.500	0.436	-	-106	-0.37

[a] Record has questionable validity; see Section V. 2.

[b] Alternative interpretations allow the values 0.250, 0.500, and 0.167, respectively. See text of Section IX. 2.

[c] Calculated using the values listed in footnote b.

TABLE IX. 2

MAGNITUDES OF LUNAR ECLIPSES OBSERVED AFTER 500

Identification		Magnitude		Sign of $\partial\mu_m/\partial\Omega$	x_2 $''/cy^2$	$\Delta\Omega$ (deg)
		Observed	Calc.			
585 Jan 21	C	0.667	0.759	-	272	0.53
590 Oct 18	C	0.775	0.784	+	- 26	-0.05
592 Aug 28	C	0.667	0.604	-	-192	-0.37
593 Aug 17	Ca	0.533	1.735	+	---	---
854 Feb 16	Is	0.833	0.925	+	-430	-0.53
856 Jun 22	Is	0.708	0.583	-	-592	-0.72
883 Jul 23	Is	0.842^b	0.950	+	-535	-0.62
901 Aug 3a	Is	0.968^b	1.066	+	-447	-0.57
901 Aug 3b	Is	0.968^b	1.066	+	-447	-0.57
923 Jun 1	Is	0.758^b	0.666	+	500	0.53
927 Sep 14	Is	0.292	0.219	+	402	0.42
979 May 14	Is	0.708	0.697	-	- 67	-0.06
979 Nov 6	Is	0.808	0.832	-	140	0.14
981 Apr 22	Is	0.250	0.182	+	414	0.39
981 Oct 16	Is	0.417	0.364	+	323	0.31
990 Apr 12	Is	0.625	0.733	-	670	0.62

[a] Record has questionable validity; see Section V. 3.

[b] 0.1 digit has been added to the numerical values listed in Table V. 4 for reasons implied there.

215

First, however, I should comment upon the observed magnitudes for the eclipses identified as -719 Mar 8 BA, -719 Sep 1 BA, and -490 Apr 25 BA. The reader will recall from Section V. 2 that the records of these eclipses, though Babylonian-Assyrian in provenance, come to us through Ptolemy [ca 152]. Ptolemy did not state whether the magnitudes in these records were of the area or the diameter, although he explicitly stated that the magnitudes in all the other records that he transmitted were of the diameter. Because of Ptolemy's remarks about older habits in measuring magnitudes (quoted in subsection V. 1. c), I tentatively concluded that the magnitudes given in these three records were of the area. On this basis, the two largest discrepancies between observed and calculated magnitudes, for any entries in either Table IX. 1 or Table IX. 2, occur for the eclipses -719 Mar 8 BA and -490 Apr 25 BA. The discrepancies become comparable with those for other eclipses if it is assumed that the recorded magnitudes are of the diameter. The calculations dealing with these eclipses will be performed on the assumption that the magnitudes are of the diameter; the correct values to use on this basis are given in Table IX. 1, footnote b. The alteration of a textual interpretation reached in Part I on the basis of numerical calculations is contrary to the spirit of this work. I do so in this case only because the alteration will have no more than a trivial effect upon the final conclusions and because it will be convenient in later discussion to have the smaller spread of the discrepancies made possible by the alteration.

I also assumed in Section V. 3, on a rather weak textual basis, that the Chinese measurements of magnitude were of the area. The results shown in Table IX. 2 for the Chinese eclipses make this assumption look reasonable.

The last two columns in Tables IX. 1 and IX. 2 are labelled x and $\Delta\Omega$; x for each eclipse is the value of $\dot{n}_M + 22''. 44/cy^2$ needed in order to make the observed and calculated magnitudes for that eclipse agree, on the assumption that Ω is given correctly by the standard ephemerides (that is, that $\Delta\Omega$ is zero). Correspondingly, $\Delta\Omega$ is the change required in Ω, on the assumption that x = 0.

The values of x listed in the tables are unreasonably large and widely scattered. The values in Table IX. 2 are much larger than those in Table IX. 1 because the dates are more recent, not because the discrepancies between observation and calculation are larger. The scatter is larger than we expect on the basis of the error estimates made in Part I, except for the Greek (M) group in Table IX. 1. The error estimates for measuring μ_m made there lead to estimated standard deviations for single values of x of about 40, 40, 120, and 70 $''/cy^2$, respectively, for the BA, M, C, and Is provenances. Unless something has been overlooked, the errors in measurement are considerably larger than I estimated in Part I, or else the node requires correcting. These possibilities will be investigated next.

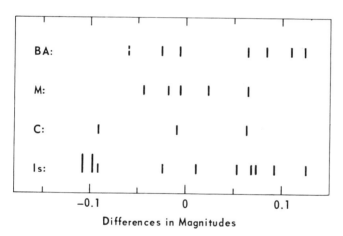

Figure IX. 1.   Distribution of the differences between observed lunar eclipse magnitudes and those calculated using standard ephemerides. A vertical stroke represents a value of the difference. Strokes corresponding to eclipse records from the same provenance appear on a horizontal line; the provenance is identified to the left of each line. The dashed stroke comes from a record of doubtful validity. The long strokes correspond to two records that happen to give the same difference. The average age of the records increases toward the top of the figure.

The distribution of the differences between the observed and calculated magnitudes listed in the tables is shown in Figure IX.1. Differences for records from the same provenance are shown on the same horizontal line. It happens that there is no overlap in time between records from different provenances, and the average age of the records decreases fairly uniformly from the top to the bottom of the figure.

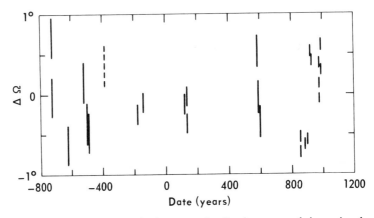

Figure IX.2. $\Delta\Omega$, which means the displacement of the node of the lunar orbit required to account for a measured lunar eclipse magnitude, as a function of the date of the eclipse. The half-length of a vertical bar corresponds to the a priori error estimate made in Chapter V. A few bars have been displaced slightly in time in order to prevent overlap. The bar drawn with dashes corresponds to a record of questionable validity.

The values of $\Delta\Omega$ listed in Tables IX.1 and IX.2 are plotted against the dates of the observations in Figure IX.2. The center point of a bar corresponds to the value listed in Table IX.1 or IX.2. The half-lengths of the bars are calculated from a priori estimates of measurement error made in Chapter V.

The M (Greek) group with dates in the -2nd and +2nd centuries are reasonably consistent with each other, exhibiting values of $\Delta\Omega$ near -0°.2. The Islamic group, between 800 and 1000, shows two clusters that have neither horizontal nor vertical overlap. Either cluster by itself would seem significant. Together, they contradict each other.

The values of $\Delta\Omega$ in Figure IX. 2 do not suggest any linear, quadratic, or harmonic variation of Ω with time that would significantly improve the agreement between the observed and calculated magnitudes in Tables IX. 1 and IX. 2. Thus we are forced to the conclusion that most of the discrepancies plotted in Figure IX. 1 are the result of measurement error and that the a priori error estimates made in Chapter V were quite optimistic.

The distributions for the different provenances shown in Figure IX. 1 do not seem to differ significantly from each other. The distribution for the BA group would extend to 0.19 rather than to about 0.12 if the original interpretation of the records had been kept, but such a difference would not be important in view of the small sample sizes and the already large scatter. The scatter of the M group is about half that for the other groups. This result is permissible for such small samples. On the other hand, Newcomb [1875; see also Section VI. 2] feared that the lunar conjunctions and occultations preserved by Ptolemy were a biased sample, and it may be that a similar fear is justified here. That is, Ptolemy may have selected the eclipse magnitudes on the basis of some type of consistency. It is unlikely that the Greek (M) measurements were as markedly superior[†] to the later Chinese and Islamic measurements

---

[†] Fotheringham [1909] studied only the BA and M groups and commented on the "superior accuracy" of the M group as compared with the BA group. He attributed this to a general improvement in astronomical knowledge and techniques in the interval between the groups. Such a conclusion would be reasonable if one had only the BA and M groups to compare. However, the results obtained from the still later Chinese and Islamic records, particularly from the latter in view of their large number, make the conclusion difficult to maintain.

as Figure IX.1 seems to indicate. I shall assume that the measurement errors have the same basic distribution for all the provenances.

There are two minor problems in the way of accepting the conclusion that the measurement errors have the distribution shown in Figure IX.1. First, the distribution looks to be more uniform in the range from about -0.1 to about 0.1 than we would expect for measurement errors; this is not a serious problem. Second, the sizes of the errors and the uniformity from one provenance to another contradict the precision used in the eclipse reports. We can only suppose that the Islamic astronomers in particular, who used a precision of $\frac{1}{2}$ digit ≈ 0.04, had not yet developed good tests for the precision of their measurements.

Figure IX.1 suggests that the error in measuring the magnitude of a partial lunar eclipse by eye has a standard deviation of about 0.07. This is much larger than the a priori estimate made in Chapter V, but it does not seem unreasonable upon further reflection. The width of the penumbra of the earth's shadow at the distance of the moon is about the same size as the diameter of the moon. Thus there is a smooth gradient of illumination across the lunar disk when the moon is partly or wholly within the penumbra. Further, the part of the moon lying within the umbra receives considerable light that has been refracted through the earth's atmosphere, so that there is no sharp gradient at the edge of the umbra. The simple experiment described in Chapter V involved high contrast and did not correspond well to a lunar eclipse.

The reports 901 Aug 3a Is and 901 Aug 3b Is are interesting. They came from Antioch and ar-Raqqah, respectively, but both were transmitted to us by Al-Battani [ca 925]. The reports agree that the eclipse was nearly but not quite total. However, the eclipse was total, with a magnitude near 1.066 (Table IX.2) unless the standard value of $\dot{n}_M$ is enormously in error. Since the edge of the umbra does not make a sharp line across the disk of the moon, it is certainly possible to mistake a total eclipse for a partial one or vice versa. However, it is peculiar that two independent observers made the same mistake and assigned the same magnitude (<1) to a total eclipse. We shall find another peculiarity of these same two reports in the next chapter.

3. SUMMARY OF RESULTS FROM LUNAR ECLIPSE MAGNITUDES

TABLE IX. 3

SUMMARY OF RESULTS FROM LUNAR ECLIPSE MAGNITUDES

Provenance	T_{eff} cy	x $"/cy^2$
BA	-25.0	-18 ± 28
M	-18.9	-46 ± 40[a]
C	-13.1	$+18 \pm 136$
Is	- 9.7	$- 6 \pm 134$

[a]Does not correspond to the observed error distribution. See text of Section IX. 3.

Results from the magnitudes of lunar eclipses are summarized in Table IX. 3, which compresses the results from each provenance into a single line. The first column of Table IX. 3 gives the provenance. The second column is called T_{eff}. Since an inferred value of x involves T^2 rather than T, I calculated the mean value of T^2 for each provenance. Let $\overline{T^2}$ denote this mean. Then $T_{eff} = - (\overline{T^2})^{\frac{1}{2}}$. The sign reminds us that the corresponding values of T are negative. The last column gives the mean value of x, taken from Tables IX. 1 and IX. 2, for each provenance, along with an estimate of a standard deviation.

Except for the M provenance, the error estimates in Table IX. 3 are based upon the scatter of the values of x about their means. The standard deviation obtained in this way for the M reports is $17"/cy^2$. This reflects the

221

smaller scatter shown for the M provenance in Figure IX. 1. Since I assumed in the last section that all provenances should have the same basic measurement errors, I based the standard deviation for the M records in Table IX. 3 upon the value for the BA records, adjusted for the differences in T_{eff} and the number of records.

The values of x are calculated on the assumption that no correction $\Delta\Omega$ is needed to the node of the lunar orbit. Because of the relation (IX. 6), the lunar magnitudes do not allow us to separate the effects of x and $\Delta\Omega$ unless the time dependence of $\Delta\Omega$ differs greatly from being quadratic in the time. Thus we cannot use the results in Table IX. 3 unless we can combine them with other measurements or with some assumption that allows separating the effects of x and $\Delta\Omega$.

In this study I shall make the simplifying assumption that $\Delta\Omega$ is zero. The reader should keep this assumption in mind when interpreting the results of the study. It will be desirable to remove the assumption in later work, either on theoretical grounds or on the basis of additional data. It will also be desirable to consider the possibility that the obliquity of the ecliptic needs modification.

Because of the large errors encountered in the measurements of lunar eclipse magnitudes, the magnitudes will not contribute much to the problem of finding the secular accelerations. This conclusion is contrary to the experience of Fotheringham [1909; also 1920, p. 124]. Fotheringham obtained apparently useful results from them because he used only magnitudes from the M provenance; these magnitudes have an unusually small scatter. For the reason explained in the second footnote in this chapter, it is not a simple matter to compare the results found here with those that Fotheringham got.

CHAPTER X

ANALYSIS OF LUNAR ECLIPSE TIMES

1. PRELIMINARY CONSIDERATIONS

The measured times of various phases of lunar eclipses were measured in solar time. This suggests that we can discuss the measured times in terms of the accelerations with respect to solar time more easily than with respect to ephemeris time. A lunar eclipse occurs only when the longitudes of the sun and moon differ by one-half revolution. The discrepancies in the mean longitudes of the sun and moon produced by the accelerations were called L and L', respectively, in Chapter I. The moon gains on the sun by the amount L - L', and the time of an eclipse is therefore advanced by $(L - L')/(n_M - n_S)$; n_S denotes the mean motion of the sun. Therefore, a measurement of a time associated with a lunar eclipse gives the difference between the accelerations called $2L/\tau^2$ and $2L'/\tau^2$ in Chapter I.

From Eqs. (I. 1) we find

$$(2L/\tau^2) - (2L'/\tau^2) = \dot{n}_M - 1.607 \times 10^9 (\dot{\omega}_e/\omega_e). \quad (X. 1)$$

Each measurement of a time associated with a lunar eclipse therefore supplies a linear relation between $\dot{n}_M$ and $\dot{\omega}_e$. The average slope of the linear relation is found by setting the right member of Eq. (X. 1) equal to a constant.

Many of the times listed in Table V. 4 were given in terms of the altitude of a celestial object. The method of finding time from a measured altitude is essentially the same whether the observation is related to a solar eclipse or to a lunar eclipse. Hence I shall discuss here the reduction of altitude to time for all observations, even though the times of the solar eclipses will not be analyzed until Chapter XII.

In finding time from a measured altitude, it is necessary to know the right ascension and declination of the celestial object used. It is not necessary to know them with high accuracy; an error of 1° in right ascension makes an error of only 4^m in time. The latitude and longitude of the moon and the longitude of the sun were taken from Nav. Obs. [1953]. The latitude of the sun was taken to be zero. Latitude and longitude were then converted to right ascension and declination. The calculations were carried to a precision of 0°.1.

The position of the sun was found for noon of the eclipse date. The maximum error produced by using this approximate time is about $\frac{1}{4}$ degree or about 1^m. The position of the moon was found for the middle time of the eclipse listed by Oppolzer [1887]. The half-duration of an eclipse is typically about 100^m. During this interval the moon moves a little less than 1°, and the time inferred is thus in error a little less than 4^m. Since the time errors are equal but opposite for the beginning and end phases, the average time error caused by approximating the moon's position is considerably less. There will be an unknown small systematic error caused by the accelerations, which make Oppolzer's times systematically wrong.

The latitude and longitude of the stars needed were found by the method of Section VIII.1, except that the proper motions were neglected because of the low accuracy requirement. Latitude and longitude were then converted to right ascension and declination. The calculations were performed with a precision of 0°.1. The right ascension and declination referred to the equinox and equator of 1950.0 were taken from Becvar [1964].

Altitude is tabulated as a function of the latitude of the observer, the declination of the object observed, and the hour angle of the object observed, in Hydro. Off. [1940]. Latitude is used with a tabular interval of 1° and declination with an interval of 0°.5. In finding a time with the aid of the tables, the tabular values of latitude and declination closest to the correct values were used, with no attempt

224

to interpolate in these variables. In a typical case the error caused by using the tabular values without interpolation is around 1^m in time. The tables in effect give single-entry tables of altitude as a function of hour angle for each tabular value of latitude and declination; hour angle was found from the altitude by inverse interpolation in the tables. Finding time from hour angle is then a trivial calculation.

In seven instances, the altitude reading and the time deduced by the ancient astronomer have both been preserved. In one instance, that of the beginning time of the lunar eclipse 925 Apr 11 Is, the discrepancy between the time deduced here from the altitude and that given in the ancient record is $1^h 43^m$. Since the time given appears accurate, as shown by the eclipse calculations, it is almost certain that either the star identity or its altitude has been copied wrong. In the remaining six instances, the algebraic mean difference between the recorded time and that deduced here is about 10^s, and the root-mean-square difference is slightly less than 10^m.

2. THE GENERAL PROCEDURE USED IN THE ANALYSIS

The method of analysis requires two times connected with each measurement. The first of these is the value of Greenwich mean solar time deduced from the ancient measurement. The second is the time calculated for the observed phase on the assumption[†] that $\dot{n}_M = -22''.44/cy^2$ and that $\dot{\omega}_e = 0$. The latter time for each measurement is taken from the eclipse calculations.

Let Δt denote the difference between the two times, in the sense Observed Time minus Calculated Time. If it

[†]This time is on the ephemeris time scale as that scale is currently used in the standard lunar ephemerides. However, it differs conceptually from ephemeris time and equals it only if the standard value of $\dot{n}_M$ is correct. For this reason I shall continue to use an awkward circumlocution to denote the kind of time in question.

were true that $\dot{n}_M = -22''.44/cy^2$ exactly, $\dot{\omega}_e$ would be given by

$$\dot{\omega}_e = 2\Delta t/T^2 \qquad (X.2)$$

in appropriate units. Define y by

$$y = 10^9(\dot{\omega}_e/\omega_e) \qquad (X.3)$$

when the units are chosen in such a way that the units of the right member are cy^{-1}. Further, let y_0 denote the value of y corresponding to the value of $\dot{\omega}_e$ given by Eq. (X.2); that is, let

$$y_0 = 2 \times 10^9 \Delta t/\omega_e T^2 . \qquad (X.4)$$

The time measurement and thence Δt contains an error. Let σ_t denote an estimate of the standard deviation of this error. A priori estimates of σ_t were made in Part I; we shall be able to make a posteriori estimates in this chapter. Let σ_y be the value calculated for the right member of Eq. (X.4) when Δt is replaced by σ_t. Then, still using the assumption that $\dot{n}_M = -22''.44/cy^2$, we can say that each measurement of a lunar eclipse time furnishes a relation involving y that has the form $y = y_0 \pm \sigma_y$.

However, we cannot assume that $\dot{n}_M = -22''.44/cy^2$, since a main purpose of the present work is to see whether the estimate of $\dot{n}_M$ can be improved. Let

$$x = \dot{n}_M + 22''.44/cy^{-2}; \qquad (X.5)$$

in this study, x will always be in units of $''/cy^2$. Then $y_0 \pm \sigma_y$ is the estimate of y if $x = 0$. From Eq. (X.1) we further see that $dy/dx = 1/1.607 = 0.622$. Therefore, each measurement furnishes a linear relation between x and y of the form

$$-0.622x + y = y_0 \pm \sigma_y . \qquad (X.6)$$

This has the form of Eq. (VII.1), which is the desired form.

A statement made in the first paragraph of this section needs a little further explanation. I stated that I took the calculated times from the results of the eclipse program. This is strictly true only when the observed phase was the middle. I wrote the eclipse program to give the times when the magnitude was zero. However, I assumed that the time corresponding to an observed beginning, for example, was the time when the magnitude reached some value $\mu_0 > 0$, on the average, and I made a similar assumption for an observed ending. For simplicity, the rest of the immediate discussion will concern only the beginning; modifications for the ending should be obvious.

Let $\dot{\mu}_0$ be the time derivative of the magnitude at the beginning of an eclipse. A time $\mu_0 / \dot{\mu}_0$ must be added to the time when the magnitude was zero. Instead of calculating $\dot{\mu}_0$ in the eclipse program, I use a rough method of estimating it. Since μ_0 is small, a rough value of $\dot{\mu}_0$ is adequate.

The assumption that the magnitude is a quadratic function of time that is symmetrical about the time of greatest eclipse leads immediately to $\dot{\mu}_0 = 2\mu_m / t_0$, in which t_0 is half the duration of the eclipse. For most of the eclipses used, t_0 lies approximately in the range from $1\frac{1}{2}$ to 2 hours, although it is shorter for a few of the smallest eclipses. Thus I adopted the approximation $\dot{\mu}_0 = \mu_m$ when time is in hours. This is crude but adequate for the present purpose.

In Section V.1, I adopted the value $\mu_0 = 0.0075$, with with a standard deviation of $\frac{1}{3}$ of this amount. The results of Chapter IX raise doubts that the adopted value is correct. However, although every discussion that I have seen of anciently measured lunar eclipse times assumes either a value of μ_0 or a constant time error, there does not seem to be any consensus about the errors to adopt. Hence I shall continue to use 0.0075 in spite of the doubts about it. Luckily, for reasons that were discussed in Section V.1, the final results of the study hardly depend upon the value used for μ_0.

3. SUMMARY OF THE RESULTS

The results obtained with the lunar eclipse times observed before the year 500 are summarized in Table X. 1, and the results obtained with observations after 500 are summarized in Table X. 2., In each table the first column gives the identification of the eclipse, including the provenance, the second column gives the phase observed, the third column gives the observed time reduced to Greenwich mean solar time, and the fourth column gives the time

TABLE X. 1

TIMES OF LUNAR ECLIPSES OBSERVED
BEFORE 500

Identification		Phase	Greenwich Mean Solar Time		Observed Minus Calc. Time[a] (hr)	y_0 cy^{-1}	σ_y^e cy^{-1}
			Observed (hr)	Calc.[a] (hr)			
- 720 Mar 19	BA	Beg.	16. 46	21. 76	-5. 30	-17. 6	0. 9
- 719 Mar 8	BA	Mid.	21. 24	25. 85	-4. 61	-15. 4	2. 1
- 719 Sep 1	BA	Beg.	15. 71	20. 82	-5. 11	-17. 0	0. 8
- 620 Apr 22	BA	Beg.	1. 62	5. 97	-4. 35	-15. 6	0. 6
- 522 Jul 16	BA	Mid.	20. 14	24. 79	-4. 65	-18. 1	2. 0
		Beg.[b]	19. 49	23. 43	-3. 94	-15. 3	1. 4
- 501 Nov 19	BA	Mid.	20. 33	25. 11	-4. 78	-18. 9	2. 3
- 490 Apr 25	BA	Mid.	20. 58	23. 77	-3. 19	-12. 8	2. 3
- 382 Dec 23	BA[c]	Mid.	3. 58	8. 72	-5. 14	-22. 5	0. 7
		Beg.[d]	3. 58	7. 95	-4. 37	-19. 1	0. 7
- 381 Jun 18	BA[c]	Mid.	17. 00	21. 85	-4. 85	-21. 3	0. 8
		Beg.[d]	17. 00	20. 55	-3. 55	-15. 6	0. 8
- 381 Dec 12	BA[c]	Beg.	18. 58	22. 02	-3. 44	-15. 1	1. 9
- 200 Sep 22	M	End	18. 43	21. 43	-3. 00	-15. 5	0. 7
- 199 Mar 19	M	Beg.	21. 47	24. 30	-2. 83	-14. 7	1. 4
- 199 Sep 12	M	Beg.	- 1. 42	1. 66	-3. 08	-16. 0	1. 4
		Mid.	0. 17	3. 47	-3. 30	-17. 1	1. 0
- 173 Apr 30	M	Beg.	22. 84	25. 43	-2. 59	-13. 8	1. 4
		End	25. 51	28. 04	-2. 53	-13. 4	0. 6
- 140 Jan 27	M	Beg.	20. 05	22. 05	-2. 00	-11. 0	1. 2
+ 125 Apr 5	M	Mid.	18. 45	20. 98	-2. 53	-18. 3	0. 9
133 May 6	M	Mid.	21. 19	23. 02	-1. 83	-13. 4	2. 0
134 Oct 20	M	Mid.	20. 75	22. 96	-2. 21	-16. 2	1. 9
136 Mar 6	M	Mid.	2. 19	3. 77	-1. 58	-11. 6	0. 8

[a]Calculated using the standard accelerations.

[b]Calculated assuming that the cuneiform record gives the time of beginning in equal hours.

[c]Will receive 0 weight; see text.

[d]Using alternate interpretation of the text.

[e]A priori estimate.

228

TABLE X. 2

TIMES OF LUNAR ECLIPSES OBSERVED
AFTER 500

Identification		Phase	Greenwich Mean Solar Time		Observed Minus Calc. Time[a]	y_0 cy^{-1}	σ_y[b] cy^{-1}
			Observed (hr)	Calc.[a] (hr)			
854 Feb 16	Is	Beg.	19.32	20.10	-0.78	-16.3	2.7
854 Aug 12	Is	Beg.	0.01	0.59	-0.58	-12.1	2.7
856 Jun 22	Is	Beg.	0.45	0.97	-0.52	-10.9	2.7
866 Nov 26	Is	Beg.	0.22[c]	1.22	-1.00	-21.4	2.8
		End	1.62[c]	1.69	-0.07	- 1.5	2.8
883 Jul 23	Is	Mid.	17.50	17.70	-0.20	- 4.4	2.9
901 Aug 3a	Is	Mid.	1.02	1.04	-0.02	- 0.5	3.0
901 Aug 3b	Is	Mid.	1.07	1.04	+0.03	+ 0.7	3.0
923 Jun 1	Is	Mid.	17.84	18.01	-0.17	- 4.1	3.1
		End	18.96	19.38	-0.42	-10.0	3.1
925 Apr 11	Is	Beg.	16.38	17.03	-0.65	-15.6	3.1
		End	19.81	20.32	-0.51	-12.2	3.1
927 Sep 14	Is	Beg.	0.92	1.64	-0.72	-17.4	3.1
929 Jan 27	Is	Beg.	20.16	21.66	-1.50	-36.3	3.2
933 Nov 5	Is	Beg.	1.37	1.88	-0.51	-12.4	3.2
979 May 14	Is	End	17.92	18.20	-0.28	- 7.5	3.5
979 Nov 6	Is	Beg.	20.09	20.52	-0.43	-11.6	3.5
		End	23.23	23.56	-0.33	- 8.9	3.5
980 May 3	Is	Beg.	---[d]				
		End	2.50	2.97	-0.47	-12.7	3.5
981 Apr 22	Is	Beg.	1.49	1.90	-0.41	-11.1	3.5
		End	3.04	3.54	-0.50	-13.5	3.5
981 Oct 16	Is	Beg.	2.22[c]	2.45	-0.23	- 6.2	3.5
983 Mar 1	Is	End	25.61	25.77	-0.16	- 4.3	3.5
990 Apr 12	Is	End	21.62	23.74	-2.12	-58.4	3.6
1001 Sep 5	Is	End	18.02	18.07	-0.05	- 1.4	3.7

[a]Calculated using the standard accelerations.

[b]A priori estimate.

[c]Will receive 0 weight; see text.

[d]Record contained an impossible condition.

calculated by the methods explained in the preceding section[†]. The fifth column gives the quantity called Δt in the preceding section, and the sixth and seventh columns give the quantities called y_0 and σ_y.

The values of σ_y were calculated from the estimates of the time errors given in Chapter V. A contribution to the time error coming from uncertainty in judging phase

---

[†]Some of the tabulated times of day are negative and some exceed 24^h. This was done in order to make the calendar dates listed be the same as those listed by Oppolzer [1887].

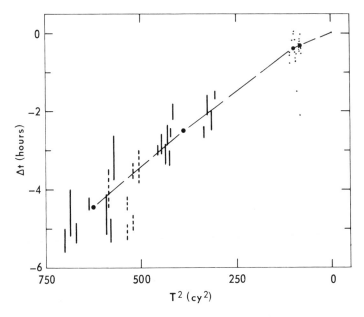

Figure X.1. Δt in hours plotted as a function of T^2 in cy^2, from measured times of lunar eclipses. Δt = observed time minus time calculated using $\dot{n}_M = -22''.44\ cy^{-2}$ and $\dot{\omega}_e = 0$. Each bar represents a single measurement. The half-length of each bar is the a priori estimate of the measurement error. The bars are not drawn for the cluster of measurements with T^2 near 100 cy^2 because they would have overlapped too much. The a priori estimate is about 0.18 hour for each point in this cluster. The dotted bars near $T^2 = 500\ cy^2$ correspond to records whose validity has been questioned.

The large circles connected by the dashed line show the mean values of Δt and of T^2 for the measurements lying within the following ranges of T^2: 750 to 500, 500 to 250, 250 to 90, and 90 to 0. The dashed line also passes through the origin.

A few bars have been displaced horizontally by a small amount in order to prevent overlap.

was included. In Part I I assumed that the uncertainty in
the magnitude at the times called beginning or end is 0.0025.
The results of Chapter IX suggest that this value is much
too small and that a better value is about 0.07 standard deviation. The corresponding contribution to the time error
depends upon $\dot{\mu}_0$ and hence varies from one eclipse to
another. It is about $0^h.15$ when averaged over all eclipses.
The change in σ_y resulting from the revision of this contribution is negligible for the eclipses before 500. For the
later eclipses the values of σ_y in Table X.2 should be multiplied by about 1.5.

The values of Δt are plotted against T^2 in Figure
X.1. A bar is used to represent each observation in Table
X.1, for which $T^2 > 250$. The half-length of a bar is the
a priori estimate of the time error. The observations for
which $T^2 > 500$ are from the BA provenance and those for
which T^2 is between 250 and 500 are from the M provenance. The cluster of points for which T^2 is near 100 represent Islamic observations. These points are too close
together to allow drawing the error bars for the observations. The standard deviation of Δt for the Islamic observations should be about $0^h.18$, including the revision of the
contribution from phase uncertainty.

T^2 is plotted as increasing toward the left in order
to make the times of the observations increase toward the
right.

The three observations represented by dashed bars
with T^2 near 500 correspond to the records -382 Dec 23
BA, -381 Jun 18 BA, and -381 Dec 12 BA. These eclipses
will be omitted from quantitative inferences for reasons
explained in Section V.2. Two bars are shown for each of
the records -382 Dec 23 BA and -381 Jun 18 BA. These
correspond to the alternate interpretations that the observed phases were the beginning and the middle. The reasons for providing alternates are given in Section V.2.

The solid bar near $T^2 = 590$, with its center at Δt
$-4^h.65$, corresponds to the Ptolemaic form of the record
-522 Jul 16 BA in which the middle phase was taken to be

1 eq. hr. before midnight. The dashed bar lying immediately to the right, with its center at $\Delta t = -3^h.94$, corresponds to the cuneiform record if the time $3\frac{1}{3}$ hours after sunset is taken to be in equal hours. The solid bar will be used in the inference of the accelerations.

Two points for which T^2 is near 100 lie well outside the cluster. These points correspond to the records 929 Jan 27 Is and 990 Apr 12 Is. The time in the latter record was given (Table V.4) as "rising of the 1st deg. of Aquarius". The discrepancy with this record is almost exactly 2^h. It is plausible that the observer meant to write "last degree of Aquarius" or, equivalently but more probably, "first degree of Pisces". The time in the record 929 Jan 27 Is was given in two ways, as "5^h un. hr. after sunset" and as "height of Arcturus 18° E". I reduced these times independently to Greenwich mean solar time and obtained agreement to the minute, so it seems unlikely that there was a copying error in the time[†]. The most likely explanation of the error is that the observed phase was actually the middle of the eclipse or possibly the beginning of totality. Whether these explanations are correct or not, it seems likely that the errors are clerical and not observational. Hence I feel justified in omitting these records from further quantitative inferences.

The coordinates of the center of the circle shown near $T^2 = 625$ are the mean values of T^2 and of Δt for the BA records. A similar mean for the M records is shown near $T^2 = 400$. Since there are more Islamic records than BA and M records combined, I divided the Islamic records into two portions, with T^2 respectively greater than and less than 90. Means are plotted in Figure X.1 for each portion. A broken dashed line connects the circles with each other and with the origin.

[†] It is possible, of course, that the record originally gave the time in only one way, that the time had been recorded wrong, and that a later editor calculated the other way of stating the time and interpolated it into the record.

If the values of $\dot{n}_M$ and $\dot{\omega}_e$ were constant, the values plotted in Figure X.1 would lie on a straight line passing through the origin, except for measurement error. As it is, the values deviate from a straight line by an amount that appears statistically significant. Thus I have not tried to find a single value of y_0 to represent all the measurements of lunar eclipse times in Tables X.1 and X.2.

TABLE X.3

SUMMARY OF RESULTS FROM LUNAR
ECLIPSE TIMES

Provenance	T_{eff} cy	Mean Δt[a] (hr)	y_0 cy^{-1}	σ_y cy^{-1}	y'
BA	-25.0	-4.47	-16.4	0.7	0.622
M	-19.8	-2.50	-14.6	0.7	0.622
Is before 950	-10.0	-0.42	- 9.6	1.8	0.622
Is after 950	- 9.2	-0.33	- 8.9	1.5	0.622

[a]Observed minus calculated Greenwich mean solar time, calculated using the standard accelerations.

The main properties of the four circles plotted in Figure X.1 are listed in Table X.3. The first column gives the provenance. The second column gives the value of T_{eff}, as this quantity was defined in the preceding chapter, and the third column gives the mean value of Δt. The fourth column gives the mean value of y_0 calculated from the values of y_0 for the individual eclipses. I shall take this value of y_0 to be the best estimate that can be formed from the measurements in the various groups. The fifth column in Table X.3, called σ_y, is an estimate of the standard deviation to be associated with the tabular y_0. The values of y_0 and T_{eff} are plotted in Figure X.2.

y' is listed in the last column for convenience in later reference, even though it is the same for all lines.

There are several ways to form the values of y_0 and σ_y listed in Table X.3 from the values listed for the individual observations in Tables X.1 and X.2. One way

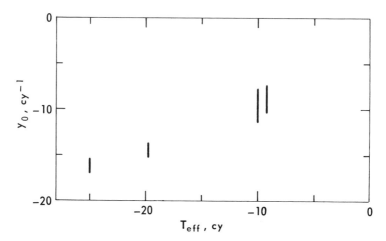

Figure X. 2. y_0 in cy^{-1} plotted as a function of T_{eff} in cy. y_0 is the average, for a group of measurements, of $10^9(\dot\omega_e/\omega_e) - 0.622$ $(\dot{n}_M + 22.44)$, and $(T_{eff})^2$ is the average of T^2 for the same group. The groups are those represented by the large circles in Figure X.1.

is to weight each observation in accordance with the a priori estimate of σ_y listed in Table X. 1 or X. 2[†]. Another way is to give the same weight to each observation. I calculated the values of y_0 for Table X. 3 in both ways and found that they differed by about 0. 1 cy^{-1}, an inconsequential amount. The values listed in Table X. 3 are those found by equal weighting. The values of σ_y in Table X. 3 are based upon the scatter of the values of y_0 for the individual eclipses about their mean, calculated separately for each grouping.

Another way was suggested in Section V. 1. a in connection with the uncertainty in judging the beginning and end phases. It was pointed out that the uncertainty did not matter, provided it was symmetrical with regard to beginning and end and provided that beginning and end were equally represented in the data. I stated the intention of testing the assumed uncertainty by: (a) weighting all phases

---

[†]Or better, in accordance with the revised estimate discussed at the beginning of this section.

equally, and (b) giving the whole set of beginning phases the same weight as the whole set of end phases. Since it now appears that the collection of observations must be divided in time because of an apparently significant time dependence, there would not be enough observations in each time division to allow a significant test.

4. DISCUSSION

The records 901 Aug 3a Is and 901 Aug 3b Is need comment. Both records were given by Al-Battani [ca 925], and they represent independent observations of the same eclipse from ar-Raqqah and Antioch. The values of Δt for the two records agree closely. However, the two values of Δt are the two algebraically largest values found for the Islamic records, and they lie about two standard deviations from the mean. Such close agreement is peculiar for values that are as seriously in error as these seem to be. The peculiarity is heightened by the recollection (see Section IX. 2) that these same two records reported as partial, and with the same magnitude, an eclipse that was in fact total.

The eclipses of -382 Dec 23 and -381 Jun 18 agree much better with the others (see Figure X. 1) if it is assumed that the reported times are for the beginning phase (upper bars) rather than for the middle phase (lower bars). Ptolemy [ca 152] adopted this interpretation, and it is a reasonable one. However, as I mentioned in Section V. 2, the adoption of this interpretation leaves a problem. If the eclipse of -382 were observed in Babylon, it began about $\frac{1}{2}$h before sunrise, more than half way from the beginning of astronomical twilight toward dawn. How then was it known that the eclipse would be a small one, with a duration estimated at only $1\frac{1}{2}$ hours?

Figure X. 1 does not suggest a significant difference between the two interpretations of the record -522 Jul 16 BA.

Figures X. 1 and X. 2 suggest a marked decrease in the quantity y_0 throughout historic times; y_0 is a certain linear combination, given by Eq. (X.6), of the accelerations $\dot{n}_M$ and $\dot{\omega}_e$. Extrapolation in Figure X.2 suggests that y_0 at the present epoch is much smaller than it was 25 centuries ago. It is not safe to reach this conclusion or to speculate about it until the results from the other forms of data have been discussed.

5. A DIGRESSION ON KEEPING UNEQUAL HOURS BY THE STARS

When I estimated the accuracy of ancient time measurements in Chapter V, I assumed that time before the Islamic period was measured by a water clock or similar device. I made this assumption mainly because all writers that I had read on the subject made it. I was also influenced by the idea of unequal hours. In the Mediterranean and Babylonian regions there seemed to be, at all periods, the freedom to use unequal hours even if equal ones were also used at times. I assumed that the use of unequal hours required a device like a marked candle or a water clock with the property of easy rate regulation.

I have realized since that unequal hours can be kept rather accurately by astronomical observations of a simple sort, and that the method could originate at an early stage in the development of an astronomy that was interested in the zodiac.

The method is mentioned by Pannekoek [1961, p. 106]. In a discussion of how the ancient Greeks determined the time, he wrote: "The progress of the hours at night was recognized by the rising of the constellations, especially those of the zodiac. Knowledge of the twelve zodiacal signs, ascribed to Oenopides of Chios (about 430 B.C.), was doubtless borrowed, as the names show, from Babylon. In the course of the night five[†] zodiacal

---

[†]Five is the number that Pannekoek wrote. If a sign or a zodiacal constellation is used to mean exactly 30°, the number should be six. "Five" could be a typographical error, or Pannekoek could have meant that only five signs would rise in their entirety on most nights.

constellations are seen to rise, beginning at nightfall with the stars opposite the sun, and ending with the stars appearing just before sunrise. So the appropriate unit of time is the double-hour, which is longer on long nights, shorter on short nights. " Pannekoek then went on to discuss how the period of daylight could be divided into the same number of portions, regardless of the season, by the use of poles or sundials.

In other words, if we neglect refraction, the flattening of the earth, and other small effects, we can say the following: Regardless of the season, and hence regardless of the length of the night, exactly half of the zodiac rises above any observer's horizon, regardless of his location[†], between sunset and sunrise. Alternatively, exactly half of the zodiac sets; however, half of it does not need to cross the meridian.

Thus, if time is kept at night by observing what part of the zodiac is rising or setting, the automatic result is to keep time at night in unequal hours[‡].

If the Babylonian and Greek astronomers did keep time this way, an immediate consequence is that the accuracy of their time measurements was uniform through the night, and that it did not depend upon how far the time was separated from sunrise or sunset, as I assumed in Chapter V.

The Babylonian units of time are suggestive but not conclusive. The longer unit was the beru or "double-hour". The shorter unit was the us or ush, equal to four of our minutes. In other words, the Babylonian units correspond to motion through one zodiacal sign and through 1° of the

---

[†] Except in the polar regions where the ecliptic can sometimes be seen in its entirety and where it can sometimes not be seen at all.

[‡] The time intervals between risings of successive signs of the zodiac are not strictly uniform, but this would probably not have bothered the earliest astronomers.

zodiac, respectively. We also know that the Babylonians used "unequal hours" for many [van der Waerden, 1956, Chap. II] non-astronomical purposes. However, the authorities are unanimous, at least within my reading, that the beru and ush were used only to keep equal hours.

On the other hand, the Greek writings that I have seen used "hour" as we use it, except for the possibility that an hour might be unequal. The Greeks did not seem to use a named smaller time unit.

In sum, according to the authorities, we have this remarkable situation: The Greeks used the "zodiacal method" of time-keeping but not the terms natural to it, while the Babylonian astronomers, in contrast, used the terms but not the method. This situation is possible, but I should like to see more discussion before accepting it as proven.

CHAPTER XI

ANALYSIS OF THE MAGNITUDES OF PARTIAL SOLAR ECLIPSES

1. GENERAL COMMENTS ABOUT THE ANALYSIS

"Magnitude", when used in this chapter without modifiers, will mean what was called "magnitude of the diameter" in Chapter V.

The magnitude μ of a solar eclipse is defined as the fraction of the solar diameter that is blocked from view by the moon. μ defined in this way is defined only when the sun is partially eclipsed; we had a similar problem with the definition of magnitude of a lunar eclipse in Chapter IX. In order to generalize the definition, imagine a plane, normal to the shadow axis, drawn through the position of an observer. Let L_1 and L_2 be the radii of the penumbra and umbra (or the central region for an annular eclipse) circles in this plane, and let m be the distance of the observer from the axis; this notation follows that used in ES, pages 245 to 246. In order to allow a common computation program for annular and total eclipses, we can adopt a sign convention that makes L_2 positive for annular eclipses and negative for total ones.

With the aid of a simple sketch, the reader should easily be able to convince himself that

$$\mu = (L_1 - m)/(L_1 + L_2) . \qquad (XI.1)$$

μ as defined by Eq. (XI.1) is a continuous function of time that agrees with the elementary definition when the sun is partially eclipsed. Hence Eq. (XI.1) furnishes the desired generalization of the earlier definition. μ has the properties listed in relation (IX.2), being greater than unity during the total phase of an eclipse and being negative when there is no eclipse.

The observations to be analyzed in this chapter are observations of the maximum magnitude μ_m reached during a partial eclipse. Therefore, the procedure to calculate the magnitude at a particular time was embedded in a program that accepts sets of general eclipse parameters (Besselian elements) for several times and searches for the time at which μ is a maximum. The program uses non-linear interpolation to find values of the eclipse elements at times other than the tabular times.

The geographical longitude of the observer is known in advance. However, the eclipse magnitude depends upon his ephemeris longitude, which in turn depends upon the acceleration $\dot{\omega}_e$ that is to be found. The first task is therefore to infer a value of the ephemeris longitude λ_{eph} from the observation of magnitude. In order to do this, I made a preliminary estimate of the ephemeris longitude of the observer, using the approximate value $10^9(\dot{\omega}_e/\omega_e) = -20$ cy^{-1}. I then calculated the maximum magnitude of the eclipse for two values of ephemeris longitude that were $10°$ apart and on opposite sides of the preliminary estimate.

Over a restricted range of ephemeris longitude, it is fairly safe to assume that μ_m is a linear function of ephemeris longitude. I found this function, for each eclipse, from the two calculations of μ_m just described. From this function it is simple to find the value λ_{eph} of ephemeris longitude that makes the calculated magnitude equal to the observed one.

Let x denote $\dot{n}_M + 22''.44$ /cy^2 as in Eq. (X.5), and let y_0 be the quantity defined in (X.6). y_0 means the estimate of $10^9(\dot{\omega}_e/\omega_e)$ formed for the value x = 0. The calculation just described, when performed with the eclipse elements found for x = 0, yields y_0 from the following formula:

$$y_0 = (151.7/T^2)(\lambda_{eph} - \lambda_{geo}) \quad \text{cy}^{-1}. \qquad (XI.2)$$

In using this, λ_{eph} and the geographical longitude[†] λ_{geo} are to be in degrees. T, as usual, means time in centuries

---

[†] In using Eq. (XI.2), longitude is to be measured positive eastward from Greenwich. The reader is warned that ES uses the opposite sign convention.

from 1900. The reader should have no difficulty in verifying Eq. (XI. 2).

In Chapter V, I made a priori estimates of the standard deviation σ_μ in the measured magnitudes. From the values of μ_m calculated for two values of λ_{eph} it is simple to estimate the derivative $d\lambda_{eph}/d\mu_m$ and thence to obtain an a priori estimate of the standard deviation σ_y of y_0.

Since B_i can be set equal to unity, all the quantities except A_i in the standard form of Eq. (VII. 1) have now been evaluated for each observation; with B_i = 1, A_i = - (dy/dx). I calculated a value of y using a value of x $\neq$ 0, for each eclipse, and evaluated dy/dx numerically.

2. SUMMARY OF THE CALCULATED RESULTS

The position of the observer obtained by using the tentative ephemeris longitude lies almost on the central path for the eclipses 891 Aug 8 Is and 1004 Jan 24 Is.

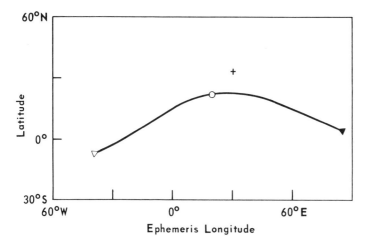

Figure XI.1. The central line of the total eclipse of 866 Jun 16. The sunrise, noon, and sunset points are marked with the symbols that Oppolzer [1887] used. The + mark is the position of the observer of the record 866 Jun 16 Is if $10^9\ (\dot\omega_e/\omega_e)$ = −20 cy^{-1}.

Since the measured magnitude for the latter eclipse (see Table V. 4) is 0. 917, a tentative position near the path is plausible. The measured magnitude for the eclipse of 891 Aug 8 was 0. 743. It seems unlikely that the correct ephemeris position is far enough from the path to make this small a magnitude possible. In other words, the measured magnitude is probably seriously in error. These two records cannot be used because there is no way to choose between the two values of y that each makes possible.

The record 866 Jun 16 Is cannot be used, for a reason shown by Figure XI. 1. The curve plotted in Figure XI. 1 is the central path of the eclipse, and the point marked + is the observer's position corresponding to the tentative ephemeris longitude. The position is far enough from the path for the measured magnitude of 0. 625 to be reasonable. However, the position is near the place where the derivative of μ_m with respect to λ_{eph} vanishes. Thus, as with the records just discussed, but for a different detailed reason, there are two possible values of y for this eclipse and there is no way to choose between them. The record 993 Aug 20 Is cannot be used for the same reason. In the latter case the tentative location is so close to the path that the magnitude almost surely cannot be as small as the value 0. 733 obtained from Table V. 4.

Ten measurements of solar eclipse magnitudes remain from the tables in Chapter V. The calculations

TABLE XI. 1

MAGNITUDES OF SOLAR ECLIPSES MEASURED
AFTER 500

Identification		Observed Magnitude	$y' = dy/dx$ $cy/''$	y_0 cy^{-1}	A priori σ_y cy^{-1}	$y_0 - 17y$ cy^{-1}
585 Aug 1	C	0.509	0.642	-17.9	2.9	-28.8
586 Dec 16	C	0.733	0.647	-19.5	2.9	-30.5
901 Jan 23a	Is	0.608	0.922	-26.3	3.0	-42.0
901 Jan 23b	Is	0.716	0.584	- 2.6	3.0	-12.5
923 Nov 11	Is	0.750	0.603	-15.4	1.6	-25.7
928 Aug 18	Is	0.367	0.186	+19.9	3.2	+16.8
977 Dec 13	Is	0.667	0.709	-17.8	1.5	-29.9
978 Jun 8	Is	0.458	0.612	- 1.2	2.6	-11.6
979 May 28	Is	0.458	0.621	- 8.6	2.2	-19.2
1221 May 23b	C	0.675	0.685	+ 4.1	10.2	- 7.5

concerning these measurements are summarized in Table XI. 1. This table is called "Magnitudes of Solar Eclipses Measured After 500"; Part I gives no magnitudes measured before 500[†]. The first column gives the identification and the second gives the measured magnitude. The next three columns give, in order, the values of $y' = dy/dx$, y_0, and the a priori estimate of σ_y.

In Part I, I assigned the value 0.04 for the standard deviation of the magnitude in the Chinese records and 0.012 for the Islamic ones. The two records of the eclipse of 901 Jan 23 suggest that the value assigned for the Islamic records is considerably too small. The two observers were so close together that the actual magnitudes at their locations cannot have differed by much more than 0.01 or 0.02, but the measurements differ by 0.108. This fact, plus the large scatter of the values of y_0, suggests that the value of σ_μ should be taken at least as large for the Islamic records as for the Chinese ones. For simplicity I decided to assign the same standard deviation σ_y to all the records. It is now necessary to decide on a value to use for this common standard deviation.

Since the values of y' differ considerably, and since the correct value of x is almost surely not zero, the scatter in the values of y_0 does not reflect the scatter in the measured magnitudes. In the absence of measurement error, the quantity $y_0 + y'x$, in which x denotes the "correct" value of $\dot{n}_M + 22''.44/cy^2$, should be constant. Hence the scatter of the values of $y_0 + y'x$ should give us a good idea of the random component of the measurement error, if not of the systematic component. The problem is to decide on the "correct" value of x.

[†] The magnitude was measured for the "eclipse of Hipparchus", which could have been any one of several eclipses, and which is discussed in Chapter IV under the listing -309 Aug 15b M. I decided in Part I to use the magnitude measurement only in an attempt to identify the eclipse and not also as an independent measurement.

I deferred this problem until the other data had been analyzed. I obtained a tentative value of x from them and decided on this basis to use $x = -17$ $''/cy^2$ in calculating $y_0 + y'x$. The values of $y_0 - 17y'$ are listed in the last column of Table XI.1. The mean of the values in this column is -19.1 cy^{-1}, and the standard deviation about the mean is 15.6 cy^{-1}. This value of σ_y corresponds approximately to a value of 0.07 for σ_μ. It should be remembered that the a posteriori estimate of σ_μ made in Chapter IX for lunar eclipse magnitudes was also 0.07.

I attempted in Chapter IX, without success, to reduce the apparent scatter of the lunar magnitudes by postulating a perturbation in the node of the lunar orbit. It may be that a similar attempt for the solar eclipse magnitudes would meet with more success. However, such an attempt would require a set of eclipse computations for perturbed values of the node and such computations are not available from the computer program. Even if they were, their use would be beyond the scope planned for this work.

Whether or not the large scatter of the values of $y_0 - 17y'$ can be explained by a perturbation in the node or by some other effect, it is reasonable to assign equal weights to the observations listed in Table XI.1 and to use a standard deviation $\sigma_y = 15.6$ cy^{-1} for all observations rather than the values listed there.

The value of about 0.07 found for the standard deviations of both lunar and solar eclipses is about what we would expect from attempts to divide an interval into equal parts by eye. This suggests that none of the ancient astronomers, not even the Islamic ones, used instrumental aids in estimating the magnitudes.

CHAPTER XII

ANALYSIS OF THE TIMES OF SOLAR ECLIPSES

1. GENERAL COMMENTS ON THE ANALYSIS

The central part of the program to analyze solar eclipse times is the procedure, described in the preceding chapter, that calculates the magnitude of a solar eclipse as a function of time. This procedure has been embedded into a program that accepts sets of Besselian elements for several times and searches for the time when the magnitude attained a pre-assigned value. Since there are two times for every magnitude less than the maximum, it is necessary to know on which side of the maximum the desired time occurred and to choose the starting times to lie on the same side.

Except with the report 1004 Jan 24 Is, the times refer to the beginning, middle, or end[†] (see Table V. 4). I decided in Chapter V to assume that the middle time is the average of the beginning and ending times, although it is plausible that middle means greatest eclipse. I also decided to take the beginning and ending times as the times when the magnitude was 0.0075. The results of Chapter IX cast considerable doubt upon this choice, at least for lunar eclipses[‡], but I decided to retain it for want of another basis for choice and because the choice, within reasonable limits, has little effect upon the final inter-ferences.

[†] Two reports give the time of greatest eclipse. Since it was not clear how the greatest eclipse was determined, I decided in Chapter V not to use these times.

[‡] The ancient astronomers probably observed solar eclipses by reflection or by an image cast by a pinhole. Either method reduces the light contrast; thus the effective contrast may have been as poor for a solar as for a lunar eclipse.

In order to calculate the ephemeris time, ET say, of a specified phase of an eclipse, we must assume a value λ_{eph} for the ephemeris longitude of the observer. Let y denote an inferred value of $10^9(\dot{\omega}_e/\omega_e)$, and let GMST denote the observed (solar) time of the specified phase. The value of y, y_λ say, that would be inferred from a value of λ_{eph} is

$$y_\lambda = (151.7/T^2)\,(\lambda_{eph} - \lambda_{geo})\quad cy^{-1}. \qquad (XII.1)$$

The right member of Eq. (XII.1) is identical with the right member of Eq. (XI.2). T is as usual the epoch of the observation in centuries from 1900, and λ_{eph} and the geographical longitude λ_{geo} should be expressed in degrees. The value of y, y_t say, that would be inferred from the times is

$$y_t = (2282/T^2)\,(GMST - ET)\quad cy^{-1}, \qquad (XII.2)$$

if the difference between the times is expressed in hours.

Since the calculated ET is a function of the assumed λ_{eph}, both y_λ and y_t are functions of λ_{eph}. The condition that y_λ and y_t should be equal therefore determines the value of λ_{eph} and hence of ET that is required by the observation. Either Eq. (XII.1) or Eq. (XII.2) then gives the corresponding value of y, from which the subscripts λ and t can now be dropped.

We are now in a position to obtain an equation of the standard form, Eq. (VII.1), for each time observation. Since y is involved in the observation, B_i is not zero and we can arbitrarily set $B_i = 1$. The calculation just described, when performed with $x = \dot{n}_M + 22.44 = 0$, yields y_0. The calculation, when performed with a different value of x, allows us to find $y' = dy/dx = -A_i$. Finally, the process of solving for the condition $y_\lambda = y_t$ yields $dy_t/d(ET)$ as a by-product; the product of this derivative by the estimated timing error is the a priori estimate of σ_y.

2. SUMMARY OF THE CALCULATED RESULTS

The observations and their comparison with the calculated results are listed in Table XII. 1. In the final inferences, observations made before 500 will be separated from those made after. Times are available for only one solar eclipse, that of 364 Jun 16, before 500. Results for that eclipse will be analyzed separately from the other eclipse times but it seems pointless to devote a separate table to them.

The double-valued nature of y, which kept us from using the recorded magnitudes of some of the eclipses in Chapter XI, does not interfere with the use of the recorded times for the same eclipses.

TABLE XII. 1

ANALYSIS OF SOLAR ECLIPSE TIMES

Identification		Phase	Observed GMST (hr)	Eph. Time Calc. with $\dot{n}_M = -22''.44/cy^2$ (hr)	y_0 cy-1	A priori Estimate of σ_y cy-1
364 Jun 16	M	Beg. [a]	12.83	14.85	-19.6	1.8
		Mid. [a]	13.80	15.73	-18.7	
		End [a]	14.50	16.54	-19.7	
829 Nov 30	Is	Beg.	4.54	4.72	- 3.6	2.6
		End	6.37	6.91	-10.7	
866 Jun 16	Is	Mid.	10.75	11.31	-11.9	2.8
		End	12.03	12.57	-11.5	
891 Aug 8	Is	Mid.	10.62	10.90	- 6.3	3.0
901 Jan 23a	Is	Mid.	6.15	6.48[b]	- 7.5	3.0
901 Jan 23b	Is	Mid.	6.12	6.54[b]	- 9.4	3.0
923 Nov 11	Is	Mid.	4.32	4.84	-12.4	3.2
		End	5.48	5.85	- 8.8	
928 Aug 18	Is	End	3.56	3.90	- 8.3	3.2
977 Dec 13	Is	Beg.	6.24	6.68	-11.9	3.6
		End	8.55	9.02	-12.7	
978 Jun 8	Is	Beg.	12.39	12.64	- 6.7	3.6
		End	14.72	15.15	-11.6	
979 May 28	Is	Sensible[c]	16.22	16.53	- 8.3	3.6
993 Aug 20	Is	Beg.	5.65	6.08	-12.0	3.7
		End	8.36	8.65	- 8.1	
1004 Jan 24	Is	0.25	14.11	14.43	- 9.1	3.8
		0.50	14.55	14.73	- 5.1	

[a]Each observation will receive a weight of 2/3.

[b]Calculated time of maximum eclipse; see Section XII. 2.

[c]Will receive 0 weight; see Section V. 1. a.

247

The first two columns in Table XII. 1 contain the identification of the eclipse record and a description of the observed phase; these columns are taken directly from Tables V. 4 and V. 5. The third column contains the observed value of Greenwich mean solar time in hours. The method by which these times were obtained from Tables V. 4 and V. 5 has already been described in connection with lunar eclipse times in Section X. 1. The final three columns contain in order the value of ET calculated for $\dot{n}_M = -22''.44/\text{cy}^2$ ($x = 0$), the value of y_0, and the a priori estimate of σ_y. The same value of σ_y is associated with each observation during a particular eclipse. This should be valid for the records left by the Islamic astronomers, who presumably measured time by measuring positions of celestial bodies. If the speculations of Chapter V, to the effect that earlier astronomers used water clocks or similar devices, are correct, the different phases in the record 364 Jun 16 M should have different a priori estimates. If the speculations of Section X. 5 are correct, the times were probably found astronomically (by a gnomon, say, for a solar eclipse) just as in the Islamic observations and should have the same error estimates.

The eclipse of 901 Jan 23 began so close to sunrise for both observers of that eclipse that there is a question whether the beginning of the eclipse could have been observed. Hence, for that eclipse only, I took "middle" to mean "maximum".

The values found for the derivative dy/dx all lie within 0. 02 from the mean value 0. 622. Hence it causes only a negligible error to take $y' = 0.622$ for the solar eclipse times after 500. This value is, as it should be, the

TABLE XII. 2

AVERAGE PARAMETERS FOR SOLAR ECLIPSE TIMES

T_{eff} cy	Mean y_0 $\text{cy}^{-1}0$	σ_y cy^{-1}	y'
-15. 4	-19. 3	1. 8	0. 614
- 9. 7	- 9. 3	0. 6	0. 622

same as the value obtained for lunar eclipse times in Chapter X. Since only one eclipse was involved before 500, I have kept the specific value of y' for that eclipse.

Since y' is nearly the same for all the observations, it is legitimate to represent a set of observations by a single value of y_0 and a single value of σ_y. The values of y_0 and σ_y, for observations before and after 500, respectively, are listed in Table XII. 2. The first line applies to the observations of the eclipse of 364 Jun 16, the so-called eclipse of Theon. The second line gives the means for the observations after 500, which are all Islamic in provenance. Since the a priori estimates of σ_y are nearly the same for all observations in this group, equal weights were used for simplicity in obtaining the values in Table XII. 2. T_{eff} is the epoch to associate with the mean. It is measured in centuries from 1900 and was calculated in the way outlined in Chapter IX.

The value of σ_y in the first line of Table XII. 2 is the same as the value associated with each observation of the eclipse of 364 Jun 16 in Table XII. 1. Although there are three times entered for this eclipse, I did this because it was decided in Chapter V to treat the times as if only two of them were independent. In assigning the value of σ_y in the second line, I first calculated the standard deviation of the Islamic values of y_0 about their mean. This standard deviation is 2. 7 cy^{-1}. This a posteriori estimate is fairly close to the a priori estimates listed in Table XII. 1, and it does not make much difference whether the standard deviation to be associated with the mean value of y_0 is based upon a priori or a posteriori estimates. I arbitrarily chose the latter.

The values of y_0 in Table XII. 2 show the same trend as the values that were obtained from lunar eclipse times and that are shown in Table X. 3 and Figure X. 2. Further discussion of this fact will be deferred to Section XIV. 5.

CHAPTER XIII

ANALYSIS OF THE LARGE SOLAR ECLIPSES

1. THE METHODS OF ANALYSIS

The immediate goal of analyzing the reports of large solar eclipses is to obtain, for each observation, a relation of the form of Eq. (VII.1) for each observation. Since the geographical path of an eclipse necessarily depends upon the acceleration $\dot{\omega}_e$ of the earth's spin, B_i cannot be zero and we can arbitrarily set it equal to unity. As usual, we shall use x to denote $\dot{n}_M + 22".44/cy^2$ and y to denote $10^9(\dot{\omega}_e/\omega_e)$ in units of cy^{-1}. We want to find values, for each large solar eclipse record, of $y' = dy/dx$, of y_0, and of σ_y.

Two methods were used in obtaining the quantities needed. Either method could be used with most records; with these the simpler method, which is basically the same as the method used by Curott [1966], was adopted. With perhaps one record out of ten, the relative geometry of the

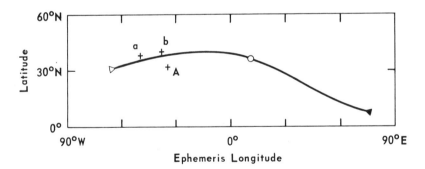

Figure XIII.1. The curve is the central line of the eclipse of −309 Aug 15 if x = $\dot{n}_M$ + 22.44 $"/cy^2$ = 0. The sunrise point, the noon point, and the sunset point are shown with the same symbols that Oppolzer [1887] used. Point a is the mean position of Agathocles allowed by the historical record, point b is the center of the Hellespont, and point A is Alexandria, if y = $10^9 (\dot{\omega}_e/\omega_e)$ = −20 cy^{-1}.

observer's position and the eclipse path required a more cumbersome method. The two methods can be illustrated with the two reports of the eclipse of -309 Aug 15.

The central line of this eclipse if x = 0 is plotted in Figure XIII. 1. The ordinate is latitude and the abscissa is ephemeris longitude; the plot is not on a Mercator projection. The observer (Agathocles) for the report -309 Aug 15a M was at latitude 37°. 5N and longitude 15°. 2E; the uncertainty in observer's position will be ignored for the moment. His ephemeris longitude calculated with the tentative value y = -20 cy^{-1} was 49°. 1W, and his corresponding position is the + mark that has "a" near it.

-309 Aug 15b M is one of the possible identifications of the eclipse of Hipparchus. The eclipse of Hipparchus, whichever eclipse it may have been, was total at "the places round the Hellespont". The position adopted for the eclipse of Hipparchus, again calculated with y = -20 cy^{-1}, is shown by the + mark with "b" near it. The eclipse was partial at Alexandria; the position of Alexandria for the same value of y is shown by the + mark with "A" near it. The position of Alexandria is shown for completeness only; the observation of magnitude at Alexandria does not enter into the discussion of this section.

The reader can see from Figure XIII. 1 that point "a" would lie on the central line if its ephemeris position were moved about 10° to the east. In order to find the change in ephemeris longitude more accurately, I plotted the relevant portion of the central line on a scale of 4° to the inch and read off the ephemeris longitude at the latitude of the observer. If $\delta\lambda_{eph}$ is used to denote the change in ephemeris longitude required to put the observer on the central line, being reckoned positive for eastward motion, the change δy in y can be written as

$$\delta y = (151. 7/T^2)\delta\lambda_{eph} \quad cy^{-1} \qquad \text{(XIII. 1)}$$

in the customary units. $-20 + \delta y$ is thus the value of y_0 inferred from the record; for the record -309 Aug 15a, $y_0 =$

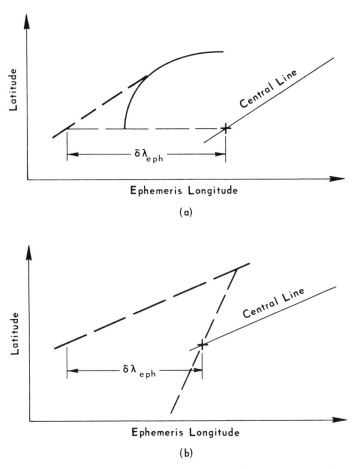

Figure XIII.2.   The uncertainty $\delta\lambda_{eph}$ in ephemeris longitude corresponding to uncertainty in observer's position, shown schematically. (a) The 1σ limit of the observer's position is the ellipse of which one quadrant is drawn. (b) The observer is at one of the ends of the slashed line through the + mark or else he is somewhere on the slashed line with 1σ limits at the ends of the line.

-16.5 cy^{-1}. This paragraph describes the simpler and more commonly used method of finding y_0.

Before going on to describe the method of finding y' and σ_y, I should point out that the value of y_0 from the record -309 Aug 15a is actually double-valued. Point "a" would also lie on the central line if $\delta\lambda_{eph} = 55°$ approximately; this would make $y_0 = -2.9$ cy^{-1}. After finding preliminary values of x and y in the manner described in Chapter VII, I had no hesitation in discarding this choice.

I found y' by repeating the procedure just described for a perturbed value of $\dot{n}_M$ and estimating y' numerically. The average value of y' must be 0.622, which is its value for lunar and solar eclipse times. However, the value of y' for an individual eclipse can depart widely from the average.

It remains to find σ_y for the "simple" method. There are three contributions to σ_y (see Chapter III).

First, there is the contribution from uncertainty in the observer's position. The cases in which the observer could be anywhere within a region are distinguished in the tables of Chapter IV by having nonzero values of the standard deviations in both latitude and longitude. The procedure used with these cases is shown in part (a) of Figure XIII. 2; this figure is schematic only and is not intended to represent a particular eclipse. I took a nominal position of the observer to be the point on the central line that has the observer's latitude. On a graph with equal scales in latitude and longitude, I drew the ellipse centered on the nominal position of the observer with semiaxes equal to the standard deviations in latitude and longitude. I then drew a curve parallel to the central line and tangent to the ellipse, and continued the tangent curve until it reached the latitude of the observer. When the difference between the longitude of this point and the longitude of the center of the ellipse is substituted for $\delta\lambda_{eph}$ in Eq. (XIII. 1), the value found for δy is the desired contribution to σ_y.

Part (b) of Figure XIII. 2 illustrates the procedure when the observer could have been at either of two points or at any point on a line. The details of the procedure should be obvious from the figure and from the preceding paragraph.

Second, there is the contribution to σ_y from the standard deviation in the magnitude given in the tables of

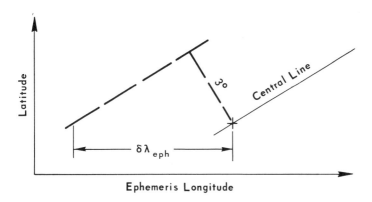

Ephemeris Longitude

Figure XIII.3. The uncertainty $\delta \lambda_{eph}$ in ephemeris longitude corresponding to an uncertainty in the magnitude of a solar eclipse, illustrated for the case in which the observer must move $3°$ normal to the central line in order to change the magnitude by one standard deviation.

Chapter IV. The procedure used in finding this contribution is illustrated in Figure XIII. 3, which is again schematic only. Since σ_y is not needed with high accuracy, and since a single contribution to it is needed with even less accuracy, a crude method of estimating the desired contribution is adequate. Assume for purposes of illustration that the standard deviation assigned to the magnitude was 0. 06. For this value I adopted a conventional displacement of 3° for the observer. That is, I assumed that the magnitude would change by 0. 06 if he moved a distance of 3°, as measured on a plot with equal scales in latitude and longitude, in a direction perpendicular to the central line. From this point I drew a curve parallel to the central line and proceeded as above. For other values of the standard deviation of the magnitude, I used a proportional value of the displacement.

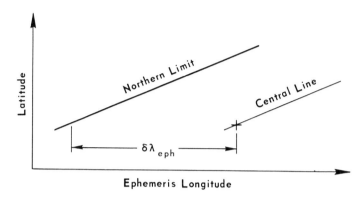

Figure XIII.4. The uncertainty $\delta\lambda_{eph}$ in ephemeris longitude corresponding to the width of the central path of a solar eclipse.

Third, there is the contribution to σ_y from the path width. The method of finding this contribution is shown schematically in Figure XIII. 4. The contribution to σ_y is the value of δy found when the value of $\delta\lambda_{eph}$ shown in the figure is substituted into Eq. (XIII. 1). In principle, the probability distribution is uniform within the path and zero outside it, and hence the result found in this way should be divided by $\sqrt{3}$. This refinement was considered unnecessary.

The resultant **a priori** estimate of σ_y is the square root of the sum of the squares of the three contributions. The estimates of σ_y need to be made using only one of the calculated paths; I chose to use the path calculated with x = 0.

The method that has been described fails with the point marked "b" in Figure XIII. 1, because the central line never attains the latitude of the observer. That is, no change in the ephemeris longitude of the observer will put him on the central line[†]. The second method, about to be described, must be used in such cases. This method also usually works more conveniently whenever the latitude of

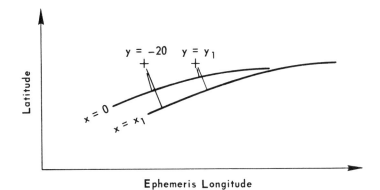

Figure XIII.5. Schematic representation of an alternate method of finding parameters associated with an eclipse record. The curves are portions of the central line calculated first with $x = \dot{n}_M + 22.44\ ''/cy^2 = 0$ and then with some other value, say x_1. The + marks are positions of the observer calculated first with the tentative value $y = 10^9\ (\dot{\omega}_e/\omega_e) = -20\ cy^{-1}$ and then with some other convenient value, say y_1.

the observer is close to the extreme latitude of the central line. The method is shown schematically in Figure XIII. 5.

The curves drawn in Figure XIII. 5 represent portions of the central line of an eclipse calculated for two values of x, say 0 and x_1. The two + marks are positions of the observer calculated for two values of y, say -20 cy^{-1} and y_1. We start by finding a value of x for each value of y.

[†] However, a change in longitude will put the observer within the zone of totality. Hence this is an allowable possibility for the eclipse of Hipparchus.

In order to find the value of x corresponding to $y = -20$ cy^{-1}, draw normals from the appropriate + mark to the two curves. Let ℓ_0 and ℓ_1 be the lengths of the normals to the curves for $x = 0$ and $x = x_1$, respectively. The reader should have no trouble convincing himself that the value of x, say x (-20), corresponding to $y = -20$ is

$$x(-20) = x_1\ell_0 / (\ell_0 - \ell_1), \qquad (XIII.2)$$

if the relation between x and ℓ is linear. Repetition of the procedure using the + mark appropriate to $y = y_1$ yields the value of x, say x (y_1), corresponding to the value $y = y_1$. From the values x (-20) and x (y_1), it is trivial to find y_0 and $y' = dy/dx$.

The procedure for finding the three contributions to σ_y and thence σ_y itself should be obvious in light of the earlier discussion.

Although the records that require the method of Figure XIII.5 are more trouble to analyze than the others, they are probably the most valuable of the large solar eclipse records. The values of y' for the records that require the second method are almost all significantly different from the mean. Hence these are the ones that contribute most to defining an intersection of the linear relations between x and y.

Two or three records required extending the central line by a small amount beyond the sunrise or sunset point in order to reach the latitude of the observer. Since the eclipse did reach near-totality in such cases, the observations are valid (unless strict totality is required) even if the path does require extension, and the uncertainty in the values of y resulting from the extrapolation is negligible.

2. RECORDS WITH AMBIGUOUS VALUES OF $\dot{\omega}_e$; REASSESSMENT OF MULTIPLE IDENTIFICATIONS

For ten eclipses the tentative position of the observer found with $y = -20$ cy^{-1} was quite close to a northernmost or southernmost point of the central line. This meant that the two values of y corresponding to each value of x were so close together that a choice between them could not be made with confidence. It also meant that the observer could be made to approach the central line either by increasing or decreasing x. The records of these eclipses had to be omitted from the inference of parameters

TABLE XIII. 1

A LIST OF ECLIPSE RECORDS THAT COULD NOT
BE USED BECAUSE THERE WAS NO WAY TO CHOOSE
BETWEEN THE BRANCHES OF A DOUBLE-VALUED FUNCTION

-688 Jan 11 M	59 Apr 30 M
-548 Jun 19 C	733 Aug 14 B
-441 Mar 11 C	809 Jul 16 B
-393 Aug 14 M	1185 May 1 E
59 Apr 30 E	1191 Jun 23 B
	1239 Jun 3 a-p E

even though the observations were apparently valid. The eclipses which were omitted for this reason are listed in Table XIII. 1.

With some of the early eclipses the difference between the geographical longitude and the ephemeris longitude of the observer is tens of degrees, and the records can yield two widely separated values of ephemeris longitude. The eclipse record -309 Aug 15a M, shown in Figure XIII. 1, is an example. The choice for these was made in the manner described in Chapter VII. It does not seem necessary to enumerate these eclipses.

The observers for the records -441 Mar 11 C and 809 Jul 16 B, which are among those listed in Table XIII. 1, must have been about 10° from the central lines, and the magnitudes must not have reached more than about 0. 8 at the locations assigned to the observers. This does not matter much for the record 809 Jul 16 B because the record

259

does not indicate totality, and an eclipse with a magnitude as small as 0.8 should certainly be observed sometimes by chance.

The record -441 Mar 11 C, however, states that stars were seen. The explanation may be that the eclipse was reported from a point outside the region assumed in Section IV.3, as Curott [1966] suggests. However, the course of the central line shows that maximum eclipse at the Chinese capital must have occurred near sunrise. This fact would make the eclipse easily noticed. It would also tend to make stars visible for a longer time into twilight or after sunrise than usual.

Since I tried not to play the "identification game" described in Section III.2, I was left with multiple possible identifications for the eclipses associated with several records. I also wrote that it seemed reasonable to drop identifications that, on the basis of detailed calculation, led to unreasonable values of the accelerations, and proposed a quantitative definition of "unreasonable".

The only tentative identification that meets the definition of unreasonableness is the possibility of 779 Aug 16 for the eclipse associated with the death of Roland (see the discussion under the listing 779 Aug 16 E in Section IV.4). The central line for the eclipse of 779 Aug 16 reaches a maximum latitude of about 41°N while the central location assigned to the observation had latitude 46°N. This record, which had to be analyzed by the method of Figure XIII.5, requires a value of x of the order of $100''/\text{cy}^2$. Accordingly, for the purposes of the present study I shall change the reliability assigned to 779 Aug 16 E from 0.005 to 0. Simultaneously I shall change the reliability assigned to the alternate 634 Jun 1 E from 0.005 to 0.01.

For the purposes of the historian, however, I do not think that the possibility of 779 Aug 16 should be dropped. The latter eclipse probably reached a magnitude between 0.85 and 0.90 at the location assigned, and there is a reasonable probability that it was seen and assimilated to the battle of 778 Aug 15. Further, the location was assigned on the basis of legend, and the correct region could have been well to the south.

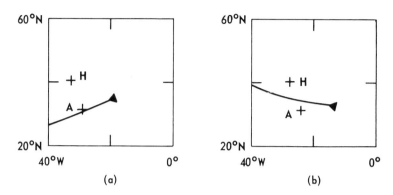

Figure XIII.6. Parts of the central lines of two eclipses, and the positions of the Hellespont (H) and of Alexandria (A) calculated with the values $\dot{n}_M = -22''.44/cy^2$ and $10^9 \times (\dot{\omega}_e/\omega_e) = -20$ cy^{-1}. The abscissa is ephemeris longitude, not geographical longitude. (a) is for the eclipse of -216 Feb 11; (b) is for the eclipse of -124 Sep 7. The solid triangles mark the sunset points.

Two tentative identifications of the "eclipse of Hipparchus" can be dropped for reasons that are shown in Figure XIII. 6. Part (a) of the figure deals with the eclipse of -216 Feb 11. The line is the central line of the eclipse, ending in the solid triangle that marks the sunset point. Points H and A are the Hellespont and Alexandria, respectively. The figure is drawn for the values x = 0 and y = -20 cy^{-1}. It can be seen that the central line fails to reach the latitude of the Hellespont, so that the eclipse cannot be made total there by any change in ephemeris longitude. Further, if we moved H to lie on the extension of the central line on the assumption that a partial eclipse so close to sunset could have been mistaken for a total eclipse, the sun would set at Alexandria before the eclipse reached its greatest magnitude. Thus it seems safe to discard the eclipse of -216 Feb 11.

Part (b) of the figure, drawn for the same values as part (a), deals with the eclipse of -124 Sep 7. If we move H and A westward together until H lies on or near the central line, A is too close to the central line and the magnitude at Alexandria is too large.

The total reliability of unity assigned to the eclipse of Hipparchus will be distributed among the four remaining possibilities. I shall assign 0.3 to each of the eclipses[†] of -309 Aug 15, -281 Aug 6, and -189 Mar 14, and 0.1 to the eclipse of -128 Nov 20.

TABLE XIII. 2

ANALYSIS OF REPORTS OF LARGE SOLAR ECLIPSES
BEFORE -400

Identification		y_0 cy^{-1}	σ_y cy^{-1}	y' $cy/''$	Relative Weight
-1062 Jul 31	BA	-16.5	0.6	0.672	0
- 762 Jun 15	BA	-18.9	2.0	0.589	0.25
- 708 Jul 17	C	-15.9	1.0	0.600	0.11
- 661 Jan 12	M	-11.7	1.8	0.576	0.003
- 660 Jun 27	M	-16.0	3.6	0.562	0.0008
- 656 Apr 15	M	-14.0	1.1	0.639	0.008
- 647 Apr 6	M	-16.8	1.8	0.568	0.003
- 634 Feb 12	M	-12.8	1.2	0.651	0.008
- 607 Feb 13	M	-21.7	7.7	0.610	0.0002
- 600 Sep 20	C	-18.5	0.9	0.644	0.14
- 587 Jul 29	M	-16.0	9.0	0.581	0.0001
- 584 May 28	M	-17.2	14.8	0.547	0.00005
- 581 Sep 21	M	- 9.9	25.0	-0.985	0.00002
- 487 Sep 1	M	-17.8	4.3	0.610	0
- 477 Feb 17a	M	-17.5	2.3	0.562	0.018
- 477 Feb 17b	M	-16.3	2.7	0.562	0
- 462 Apr 30	M	-20.2	4.8	0.536	0
- 430 Aug 3a	M	-12.4	1.3	0.658	0.61
- 430 Aug 3b	M	-11.4	2.0	0.651	0.026

Note: The weighted mean date of the reports in this table is about -700.

[†] Although he used -128 Nov 20 as the correct identification, Fotheringham [1920, p. 125] stated that the date -309 Aug 15 also satisfies the record. He did not investigate any other possibilities.

TABLE XIII. 3

ANALYSIS OF REPORTS OF LARGE SOLAR ECLIPSES
BETWEEN -400 AND +60

Identification			y_0 cy^{-1}	σ_y cy^{-1}	y' $cy/''$	Relative Weight
-	381 Jul 3	C	-22.4	2.4	0.647	0.018
-	363 Jul 13	M	-26.0	3.6	0.594	0.008
-	309 Aug 15a	M	-16.5	2.0	0.703	0.025
-	309 Aug 15b	M	+ 3.2	8.9	2.50	0.0038
-	299 Jul 26	C	-18.9	3.4	0.589	0.009
-	281 Aug 6	M	-29.7	1.0	0.532	0.30
-	189 Mar 14	M	-14.5	0.8	0.619	0.51
-	187 Jul 17	C	-19.9	2.4	0.638	0.18
-	180 Mar 4	C	-17.5	3.3	0.554	0.091
-	146 Nov 10	C	- 2.7	6.1	0.943	0.027
-	128 Nov 20	M	-15.9	0.7	0.657	0.22
-	88 Sep 29	C	-23.4	4.4	0.599	0.052
-	79 Sep 20	C	-13.2	2.3	0.651	0.19
-	50 Mar 7	E	-18.9	4.5	0.582	0
-	27 Jun 19	C	-14.2	3.5	0.654	0.080
+	2 Nov 23	C	-15.7	3.6	0.603	0.078

Note: The weighted mean date of the reports in this table is about -140.

TABLE XIII. 4

ANALYSIS OF REPORTS OF LARGE SOLAR ECLIPSES
BETWEEN +60 AND +500

Identification		y_0 cy^{-1}	σ_y cy^{-1}	y' $cy/''$	Relative Weight
65 Dec 16	C	-14.5	3.3	0.614	0.009
120 Jan 18	C	-16.0	4.1	0.535	0.006
243 Jun 5	C	-37.5	5.0	0.606	0.004
360 Aug 28	C	-33.4	25.7	1.487	0.0002
360 Aug 28	M	- 5.2	3.1	0.623	0.054
402 Nov 11	E	- 7.5	5.0	0.700	0.041
418 Jul 19	E	-20.0	11.3	0.633	0.008
418 Jul 19	M	-21.0	2.6	0.510	0.14
429 Dec 12	C	-20.8	2.8	0.083	0.013
447 Dec 23	E	-18.3	5.2	0.708	0.038
484 Jan 14	M	-14.0	2.6	0.536	0.076

Note: The weighted mean date of the reports in this table is about 420.

3. SUMMARY OF THE RESULTS

TABLE XIII. 5

ANALYSIS OF REPORTS OF LARGE SOLAR ECLIPSES
AFTER 500

Identification		y_0 cy^{-1}	σ_y cy^{-1}	y' $cy/"$	Relative Weight
538 Feb 15	E	+ 7.7	9.0	0.640	0.012
540 Jun 20	E	+ 0.5	10.1	0.702	0.010
590 Oct 4	M	-14.2	4.2	0.656	0.056
634 Jun 1	E	-21.0	14.4	0.687	0.00005
664 May 1	B	- 6.2	17.2	0.556	0.0034
878 Oct 29	B	-14.2	10.8	0.557	0.0085
1030 Aug 31	E	- 2.6	18.6	0.574	0.0015
1133 Aug 2a	B	+ 7.9	9.8	0.656	0.0053
1133 Aug 2b	B	+16.1	7.9	0.673	0.008
1140 Mar 20a	B	-26.0	15.4	0.595	0.0042
1140 Mar 20b	B	-23.2	12.2	0.608	0.0034
1221 May 23a	C	+15.1	24.8	0.623	0.0016
1241 Oct 6	E[a]	- 9.1	2.2	0.635	0.20

[a]This entry represents 14 independent reports.

Note: The weighted mean date of the reports in this table is about 1050.

The parameters obtained from the various records
of large solar eclipses are summarized in Tables XIII. 2
through XIII. 5. Table XIII. 5 contains all the records after
500. The records before 500 are presented in three groups
for convenience; the groups are those before -400, those
between -400 and +60, and those between +60 and 500. The
dividing epochs -400 and +60 are purely arbitrary.

The first column in each table gives the identification. The next three columns give the values of y_0, σ_y,
and y' needed for the standard equation. The last column
is the relative weight that will be assigned to each record.
The relative weight is the "reliability", as defined in Section III. 1, and tabulated for each record in Chapter IV,
divided by the square of σ_y.

Since the values of y' and of the relative weight vary
widely from one record to another, I do not see any particularly significant way to find <u>a posteriori</u> values of σ_y. All
one can say is that the values of σ_y appear upon inspection
to be reasonably consistent with the scatter in y_0, although
there are a few cases in which the value assigned to σ_y
should perhaps have been larger.

264

The values of y' differ enough for the records before 500 to allow a solution for x and y from the large eclipses alone. The results are

$$\dot{n}_M = x - 22.44 = -41.8 \pm 5.3 \; ''/cy^2,$$

$$10^9 \, (\dot{\omega}_e/\omega_e) = y = -27.7 \pm 3.4 \; cy^{-1}. \qquad \text{(XIII. 3)}$$

These agree reasonably well with the values found in Section II. 5 from observations of solar and lunar position.

There is no unique way to assign an epoch to the values in Eq. (XIII. 3). In earlier cases, when we were dealing with observations having about the same value of y', it was reasonable to find a mean value of the square of the age of an observation and to use the square root of the mean as the effective age of a group of observations. Here, the derivative y' is as significant as the age. For simplicity, I shall use the mean calendar date, using the weights listed in the tables in finding the mean. The weighted mean dates for the individual tables are given in notes at the foot of each table. The epoch to associate with the values in Eqs. (XIII. 3) is about -360.

CHAPTER XIV

THE ACCELERATIONS AND THEIR VARIATIONS
IN TIME

1. RECAPITULATION OF THE MATERIAL AVAILABLE FOR ESTIMATING THE ACCELERATIONS

The goal of the analysis in the chapters immediately preceding has been to find, for each observation, the parameters in the standard form of Eq. (VII. 1). It is useful to remind the reader of this equation and of the notation that has been used in recent chapters in connection with it. Instead of $\dot{n}_M$ and $\dot{\omega}_e$, the following quantities have been used:

$$x = \dot{n}_M + 22''.44 \, / cy^2,$$
$$y = 10^9 (\dot{\omega}_e / \omega_e) \quad cy^{-1}. \qquad (XIV. 1)$$

x and y will be used only when the units are those given in Eqs. (XIV. 1). In terms of x and y, Eq. (VII. 1) takes the form

$$A_i x + B_i y = Z_i \pm \sigma_{Z_i} \qquad (XIV. 2)$$

for the ith observation. Any parameter in Eq. (XIV. 2) that is not zero can be chosen arbitrarily. The convention used has been to choose $B_i = 1$ whenever it does not vanish. With this convention the B_i are either 0 or 1.

The A_i can be calculated with high accuracy, and it is legitimate to assume that all observational errors are represented by the σ_{Z_i}.

To high accuracy the A_i are the same for each observation within a given class for a number of classes. For these classes, I have gathered together all observations made within a convenient time span, or within a given

TABLE XIV. 1

MATERIAL AVAILABLE FOR ESTIMATING THE ACCELERATIONS
OTHER THAN MAGNITUDES OF PARTIAL SOLAR ECLIPSES
AND OBSERVATIONS OF LARGE SOLAR ECLIPSES

Type of Observation	Number of Table or Equation in Text	Approx. Epoch (yr)	$A_i^{\,b}$	$B_i^{\,b}$	$Z_i^{\,b}$	$\sigma_{Z_i}^{\,b}$
Babylonian lunar eclipse times	X. 3	-600	-0. 622	1	-16. 4	0. 7
Babylonian lunar eclipse magnitudes	IX. 3	-600	1	0	-40. 4	28. 0
Hipparchus' equinoxes	(II. 6)	-140	0	1	-23. 0	5. 0
Lunar occultations and conjunctions	(VIII. 1)	-100	-0. 576	1	-17. 9	1. 1
Mediterranean lunar eclipse times	X. 3	- 80	-0. 622	1	-14. 6	0. 7
Mediterranean lunar eclipse magnitudes	IX. 3	+ 10	1	0	-68. 4	40. 0
Theon's solar eclipse times	XII. 2	360	-0. 614	1	-19. 3	1. 8
Chinese lunar eclipse magnitudes	IX. 3	590	1	0	- 4. 4	136. 0
Islamic equinoxes and solstices	(II. 8)	840	0	1	-17. 0	6. 0
Islamic lunar eclipse times before 950	X. 3	900	-0. 622	1	- 9. 6	1. 8
Islamic lunar eclipse magnitudes	IX. 3	930	1	0	-28. 4	134. 0
Islamic solar eclipse times	XII. 2	930	-0. 622	1	- 9. 3	0. 6
Venus occultations and conjunctions	(VIII. 7)	930	0	1	+20. 0	24. 0
Islamic lunar eclipse times after 950	X. 3	980	-0. 622	1	- 8. 9	1. 5
L_S, Hakemite tables	(II. 11)	1000	0	1	-25. 1	5. 6
L_M, Hakemite tables	(II. 12)	1000	-0. 576	1	-11. 6	0. 4

[a]A number in parentheses is an equation number; a number without parentheses refers to a table.

[b]The units in these columns vary from line to line and are not given. Each entry is in the standard set described in Section XIV. 1.

provenance, and have represented the collection of observations by a single set of parameters for Eq. (XIV. 2).

The sets of parameters that have been formed in
this way are listed in Table XIV. 1. The first column describes briefly the class of observation. The second column
gives the place in the body of this work where the parameters were first given. The identifier in this column refers
to an equation if it is placed within parentheses and refers

to a table if it is not. The entries are arranged chronologically in order of their approximate effective epochs as listed in the third column. The remaining columns give the parameters for use in Eq. (XIV. 2).

The A_i vary widely from one observation to another for the measurements of magnitudes of partial solar eclipses and for the observations of places where solar eclipses were nearly total. Valuable information would be lost if average values of the A_i were used for these classes. Hence a separate equation of the form of Eq. (XIV. 2) will be used for each observation of these two classes. The parameters in the equations are listed in Table XI. 1 and in Tables XIII. 2 through XIII. 5. For each observation listed in these tables, the entry under y' is the negative of A_i for that observation, and the entry under y_0 is the value of Z_i. In Table XI. 1 the entry under σ_y should be used as the value of σ_{Z_i}, but this is not so for the tables dealing with the

large solar eclipses. The quantity listed as σ_y for a large solar eclipse was used only in finding the relative weight, which depends upon an assessment of the textual material as well as upon σ_y, the a priori estimate of the errors in observation. To find σ_{Z_i} for the large solar eclipses, let W_i denote the quantity tabulated under "Relative Weight" in the tables. Then

$$\sigma_{Z_i} = 1 / \sqrt{W_i} . \qquad\qquad (XIV. 3)$$

All values of Z_i and of σ_{Z_i} have been given to the first decimal place. This precision is warranted for some of the values. It has been used for consistency for all of them, even though the size of σ_{Z_i} makes this precision absurd for others.

2. THE METHOD OF INFERENCE

The method to be used for inferring the accelerations will be the method of weighted least-squares. A brief

269

description of the method will be given, principally as an aid in defining the notation.

The Z_i, the "measured quantities", have a variety of units and standard deviations. The simplest procedure in such a case is probably to observe that measurements are never perfect and hence that σ_{Z_i} never vanishes. Division of Eq. (XIV. 2) by σ_{Z_i} yields

$$a_i x + b_i y = z_i \pm 1 , \qquad (XIV. 4)$$

in which the meanings of the a_i, b_i, and z_i are obvious. All the "measurements" z_i now have the same estimated standard deviation, namely unity, and all equations of the form of Eq. (XIV. 4) receive the same weight.

Let r_i, the ith residual, be defined by

$$r_i = z_i - a_i x - b_i y ; \qquad (XIV. 5)$$

some writers use the opposite sign convention from that in Eq. (XIV. 5). Each r_i is a function of the unknowns x and y. The "best estimates" of x and y are taken to be those that minimize the function $F(x, y)$ defined by

$$F(x, y) = \tfrac{1}{2} \Sigma_i r_i^2 . \qquad (XIV. 6)$$

The sum over i in Eq. (XIV. 6) is to extend over all observations that are to be used in the inference.

(x, y) considered as a vector is the solution of

$$\begin{pmatrix} \Sigma a_i^2 & \Sigma a_i b_i \\ \Sigma a_i b_i & \Sigma b_i^2 \end{pmatrix} \begin{pmatrix} x \\ y \end{pmatrix} = \begin{pmatrix} \Sigma a_i z_i \\ \Sigma b_i z_i \end{pmatrix} . \qquad (XIV. 7)$$

Let $\underline{M}$ denote the inverse to the matrix in the left member of Eq. (XIV. 7), and let M_{xx}, and so on, denote its coefficients. Then

$$\begin{pmatrix} x \\ y \end{pmatrix} = \begin{pmatrix} M_{xx} & M_{xy} \\ M_{yx} & M_{yy} \end{pmatrix} \begin{pmatrix} \Sigma a_i z_i \\ \Sigma b_i z_i \end{pmatrix} . \qquad (XIV.8)$$

$\underline{\underline{M}}$ is clearly a symmetric matrix.

If the original estimates of the σ_{z_i} were wisely chosen, so that unity is a valid estimate of the standard deviation of each z_i (Eq. (XIV.4)), the standard deviation of the r_i calculated using x and y from Eq. (XIV.8) should also be unity. It is useful to test the validity of the weights assigned to the observations by calculating the standard deviation of the r_i and comparing it with unity.

If $\sigma(r_i)$ is the standard deviation of the r_i, the standard deviations of the best estimates of x and y from Eq. (XIV.8) are

$$\sigma_x = \sigma(r_i)\sqrt{M_{xx}} , \qquad \sigma_y = \sigma(r_i)\sqrt{M_{yy}} , \qquad (XIV.9)$$

respectively. If $\sigma(r_i)$ is unity as it should be, σ_x and σ_y reduce simply to $\sqrt{M_{xx}}$ and $\sqrt{M_{yy}}$.

TABLE XIV. 2

SOME INTERFERENCES OF THE ACCELERATIONS

Data Sample	$\dot{n}_M$ "/cy^2	$10^9 (\dot{\omega}_e/\omega_e)$ cy-1	Stand. Dev. of "r" Residuals[a]
All observations before 500	-41.6 ± 4.3	-27.7 ± 3.4	0.90
All large solar eclipses before 500	-41.8 ± 5.3	-27.7 ± 3.4	0.84
All observations before 500 except "eclipse of Hipparchus"	-41.8 ± 4.8[b]	-27.9 ± 3.0[b]	----[c]
All observations after 500	-42.3 ± 6.1	-22.5 ± 3.6	1.01
Above, but with the weights assigned to solar eclipse magnitudes multiplied by 5	-32.8 ± 5.7[b]	-17.1 ± 3.4[b]	----[c]
All observations after 500 except "Hakemite L_S"	-39.3 ± 7.8[b]	-20.7 ± 4.7[b]	----[c]

[a] This standard deviation should be unity if the data weights have been assigned accurately.

[b] Based upon the assumption that the standard deviation of the residuals is unity. These error estimates cannot be compared exactly with the others.

[c] Not calculated.

3. THE INFERRED ACCELERATIONS AND THEIR STABILITY

The accelerations $\dot{n}_M$ and $10^9(\dot{\omega}_e/\omega_e)$, which are related to x and y by Eq. (XIV.1), will first be inferred using, respectively, all observations before 500 and all observations since 500. The results are listed in Table XIV.2. For the observations before 500, we get

$$\dot{n}_M = -41.6 \pm 4.3''/cy^2, \quad 10^9(\dot{\omega}_e/\omega_e) = -27.7 \pm 3.4 \ cy^{-1} .$$

$$(XIV.10)$$

We can assign -200 as an approximate epoch at which the values in Eq. (XIV.10) apply. For the observations after 500, we get

$$\dot{n}_M = -42.3 \pm 6.1''/cy^2, \quad 10^9(\dot{\omega}_e/\omega_e) = -22.5 \pm 3.6 \ cy^{-1}$$

$$(XIV.11)$$

We can assign +1000 as an approximate epoch at which the values in Eq. (XIV.11) apply.

The calculated values of $\sigma(r_i)$ are listed in the last column of Table XIV.2. They are rather close to unity and hence the assignments of the σ_{z_i} are probably reasonable.

Estimates of standard deviations made in this paper reflect mostly estimates of precision or "scatter". It is well known that such estimates tend to be small; in other words, the error in a measurement has a strong tendency to be larger than the measurer's estimate. It is never safe to rely upon estimates of errors that result, like those quoted, from the application of equations like Eq. (XIV.9). Another way to estimate the quality of a parameter inference is to test the stability of the inferred values against reasonable changes in the data sample used.

An extremely large change in the data sample before 500 comes from leaving out all data except observations of large eclipses, or, in other words, to use only large solar eclipses. This yields the inference performed in the preceding chapter. The results are given in Eqs. (XIII. 3) and are repeated in Table XIV. 2 for convenience. This result is significant because the large eclipses contribute only about one-third of the total weight of all observations before 500. That is, about 2/3 of the total data by weight, including all data of many different classes, has been omitted in going from the first to the second line in Table XIV. 2. The differences between the two lines are trivial.

The "eclipse of Hipparchus", discussed under the listing -309 Aug 15b M, is one of the eclipses that cannot be identified from the record. It is the only such eclipse for which the possible identifications received more than a trifling weight. It is questionable whether the procedure that I followed with unidentified eclipses, which was to distribute the total weight assigned to a record among several possible identifications, is valid since the procedure cannot avoid all aspects of the "identification game".

It is desirable to test whether the method used for multiple identifications has biased the inferences. Since the only multiple identifications with appreciable weight are those for the eclipse of Hipparchus, it is sufficient to make the test by omitting all possibilities for that eclipse. The results are shown in the third line of Table XIV. 2. The changes made in the "best estimates" of $\dot{n}_M$ and $10^9(\dot{\omega}_e/\omega_e)$ are trivial. The estimates of the standard deviations are slightly different. Since the estimates were made by using unity for $\sigma(r_i)$ rather than an explicitly computed value, the differences may or may not be significant.

The measurements of solar eclipse magnitudes were all made after 500. In Chapter V, I assigned an error estimate to these measurements on the basis of the precision used by the observers. As a result of the detailed study of the measurements made in Chapter XI, I increased the error estimate considerably and used the larger estimate in the inference called "all observations after 500" in Table XIV. 2. If I had used the original estimate, the solar eclipse magnitudes would have received about five times the weight that they actually did.

It will appear in a later section that the solar eclipse magnitudes apparently have the largest numerical bias from the mean of all observations, at least of all observations that involve both $\dot{n}_M$ and $\dot{\omega}_e$. Therefore I performed a test in which I increased the weight assigned to each measurement of a solar eclipse magnitude by a factor of 5. The results are shown in the fifth line of Table XIV. 2. The values of $\dot{n}_M$ and of $10^9(\dot{\omega}_e/\omega_e)$ are changed by about 1.5 times their estimated standard deviations. This is the only test in which I could find a significant change in the accelerations, and the data weights involved in this test seem unreasonable.

The values of L_S and L_M, which are Islamic values of the mean longitudes of the sun and moon at the epoch 1000 Nov 30 at noon Cairo mean solar time, must have been derived from many observations. It is possible that the observations used [van der Waerden, 1969] to find L_S and L_M included observations, such as the eclipse times listed in Table V. 4, that have been used elsewhere in this study. It seems certain that L_S and L_M involved other observations as well.

Thus it would be an error not to use L_S and L_M because to omit them would be to throw away information. Using them means that I effectively doubled the weight assigned to those observations, if any, that have been used independently here and that were also used in estimating L_S and L_M. Since the weights are necessarily rather rough estimates, it is unlikely that any serious consequences follow from inadvertently doubling the weights attached to a small fraction of the total observations; doubling the weight of an observation corresponds to changing the estimate of its standard deviation by about 40 percent.

In order to test the possibility of serious consequences, I omitted the "observation" L_S from the set of observations after 500. As the last line of Table XIV. 2 shows, the omission causes an insignificant change in the inferred accelerations.

In summary, the accelerations inferred in this section seem to be stable against reasonable changes in the data sample. Only one data change tested resulted in a parameter change comparable with the estimates of the standard deviations. This change, which was to quintuple the weights assigned to measurements of solar eclipse magnitudes, seems unreasonable in the light of the detailed study of magnitude measurements made in Chapter XI.

4. COMPARISON WITH FOTHERINGHAM'S RESULTS

Fotheringham's study [Fotheringham, 1920] contains the most famous attempt to derive both $\dot{n}_M$ and $\dot{\omega}_e$ from ancient data alone. It is probably the study that most people have in mind when they say that there is evidence for larger accelerations in the past than now. A comparison with Fotheringham's results is thus interesting; it is also slightly complex.

TABLE XIV. 3

FOTHERINGHAM'S RESULTS AND A COMPARISON
WITH THEM

Data Sample[a]	L_{F_2} $"/cy^2$	L'_{F_2} $"/cy^2$	$\dot{n}_{M_2}$ $"/cy^2$	$10^9(\dot{\omega}_e/\omega_e)$ cy^{-1}
Solar eclipses only, F	10.55[b]	1.31[b]	-26.1	-20.2
All except eclipses, F	10.61[d]	2.04[d]	-45.5[d]	-31.4[d]
All, F	10.8	1.5	-30.7	-23.1
All except eclipses, F, corrected	10.58	2.18	-49.4	-33.6
All, F, corrected	10.6?[c]	1.6?[c]	-34 ?[c]	-25 ?[c]
All before 500, II	-----	-----	-41.6 ± 4.3	-27.7 ± 3.4

[a] F refers to Fotheringham's paper; II refers to Part II of the present study.

[b] Fotheringham never obtained values from solar eclipses only, but these values are implied by his procedure.

[c] It is not certain how Fotheringham would have combined the eclipse data with other data, so I have placed "?" after these numbers. He would certainly have gotten numbers close to these.

[d] These are the numbers Fotheringham [1920] gave for the solution of his system of relations on page 124. I get 10.58, 1.91, -42.1, and -29.4, respectively. The differences are not large.

275

Contrary to a common opinion, Fotheringham did not obtain estimates of the accelerations from the solar eclipses, although a specific estimate is implicit in his work, and he was only one step away from making it at one point. The values implied by Fotheringham's treatment of eclipses are given in the first line of Table XIV. 3. The quantity L_F', which Fotheringham called L', is identical with the quantity also called L' in Eqs. (I. 1). The quantity L_F, which Fotheringham called L, is not the same as L in Eqs. (I. 1); instead, the quantity called L in Eqs. (I. 1) equals L_F - 6.1 ($''/cy^2$). Fotheringham did not quote the quantities $\dot{n}_M$ and $10^9(\dot{\omega}_e/\omega_e)$. The latter quantities, when quoted in connection with Fotheringham's results, are calculated from Eqs. (I. 1). Fotheringham decided that a certain region in the L_F - L_F' plane was allowable on the basis of the eclipses. The values of L_F and L_F' listed in the first line of Table XIV. 3 are the coordinates of the centroid of this region.

Fotheringham did not consider the possibility that the accelerations might vary with time, and hence he did not associate an epoch with his results. His effective epoch is somewhere around 0, and his results can be compared with the first line in Table XIV. 2. The values of $\dot{n}_M$ and $10^9(\dot{\omega}_e/\omega_e)$ from that line, which are derived from all observations before 500, are repeated in the last line of Table XIV. 3 for convenience.

The accelerations implied by Fotheringham's treatment of eclipses are much smaller in magnitude than the accelerations found in this study. This is probably a consequence of the fact that most of the eclipse "identifications" used by Fotheringham came from playing the identification game. As we saw in Chapter III, the identification game is inherently conservative. That is, it tends to return to the player the same accelerations that he started with; for Fotheringham, the starting accelerations were small. He moved in the correct direction from his starting values probably because a few of his eclipse data are genuine.

The value of $\dot{n}_M$ from Fotheringham's eclipses alone is close to the "modern" value of $-22''.44/cy^2$. If only his eclipses had entered the problem, the idea of a difference between ancient and modern values of the accelerations might never have arisen and this study might never have been done.

However, Fotheringham [1920] used the large solar eclipses only after he had used a considerable amount of other data. Altogether, he used the lunar occultations from the Almagest (also used here in Section VIII. 2), Hipparchus' equinoxes (also used in Section II. 1), and the Greek but not the Babylonian measurements of lunar eclipse magnitudes and times from the Almagest (see Chapter IX). Before using the large solar eclipses at all, he formed a best estimate from this collection of data. If he had cor- rectly solved the correct equations of condition for the col- lection, he would have obtained values close[†] to those given in the line in Table XIV. 3 called "All Except Eclipses, F, Corrected". If he had found the correct solution to the equations of condition that he quoted [Fotheringham, 1920, p. 124], he would have found the values given in Table XIV. 3, footnote d. These values are remarkably close to those found in the preceding section. Further, if he had used the correct functional dependence of the lunar eclipse magnitudes and the estimate of their accuracy found in Chapter IX, he would have got $\dot{n}_M = -43".0/cy^2$ and $10^9(\dot{\omega}_e/\omega_e) = -29.9 \; cy^{-1}$, which are also close to the values found here.

The values he actually found from all the data ex- cept the solar eclipses are given in the second line of Table XIV. 3. These differ from the values given in the footnote by less than one standard deviation. It may be that

---

[†] The main error that Fotheringham made was to assume (see Section IX. 1) that lunar eclipse magnitudes depend upon the "acceleration of the sun" but not upon the "accel- eration of the moon". It is not possible to say exactly what he would have got from the magnitudes if he had not made this error, because he also attributed an abnormally small standard deviation to the magnitude measurements (Section IX. 3). In Table XIV. 3 I have guessed that he would have found the same ratio of standard deviation to mean value regardless of his method of analysis.

Fotheringham approximated the solution of his equations of condition because of the size of the standard deviations. The values that he found from all data including the eclipses are given in the third line of the table. He did not use a rigorous procedure in combining the eclipse results with the other results. His procedure had approximately, but not exactly, the effect of giving the eclipses about twice the weight of the other observations.

In summary, the values that Fotheringham found from the data other than the large eclipses are close to those found in this study for the approximate epoch -200. If he had analyzed these data correctly, his values would have been remarkably close. The accelerations that he found when he included the eclipses are much smaller in magnitude. Since Fotheringham did not develop a quantitative procedure for dealing with the large solar eclipses, it is not possible to give a rigorous explanation of the reason. However, the direction of the change made by his introduction of the eclipses is that which we would expect from his extensive use of the logical fallacy that I have called the identification game.

5. A PROBABLE TIME HISTORY OF THE ACCELERATIONS

The pairs of accelerations found from the data before and after 500, which are exhibited in Table XIV. 2, do not differ significantly. If we had no other information, we should probably conclude that the accelerations have been constant during historical time. Two independent lines of additional investigation suggest that this conclusion is wrong and suggest almost identical time histories of $\dot{n}_M$.

Spencer Jones [1939] found the value $\dot{n}_M = -22.44 \pm 1.24$ $''/cy^2$, which is the current standard, by studying lunar and planetary observations[†] from about 1750 to about

---

[†]Although it is customary to say that Spencer Jones found $\dot{n}_M$, the acceleration of the moon on an ephemeris time base, he did not write in such terms. He wrote in terms of

1930. It is reasonable to assign the epoch 1840 to his value. Newton [1968a] found $\dot{n}_M$ = -20.1 ± 2.6 $''/cy^2$ from a study[†] of the orbits of near-earth satellites. The epoch of his value is 1964. The values given in Table XIV.2 for the epochs -200 and +1000 are -41.6 ± 4.3 and -41.8 ± 6.1 $''/cy^2$, respectively. It took considerable alteration of the data sample to move the latter value to -32.8, and no possibility tested moved the former value appreciably. Thus there is a strong presumption that $\dot{n}_M$ has changed by a factor of about 2 within historic times.

The four values found for $\dot{n}_M$ are plotted against their respective epochs in Figure XIV.1. A linear variation of $\dot{n}_M$ with time can certainly not be ruled out. However, a quadratic variation is more probable, and an independent consideration to be described later strongly suggests a quadratic or higher order variation. Therefore I suggest using the parabola plotted as the dashed curve in Figure XIV.1 as a probable variation of $\dot{n}_M$ with historic time, until more data are analyzed. The parabola was not fitted to the data. It was simply calculated to pass through the values -22 at 1900, -42.5 at 1000, and -41 at -200[‡]. The equation of the parabola, along with estimates of the uncertainties in its parameters, is:

---

the "accelerations of the sun and moon", and he had to use results from ancient observations in order to find them. He exhibited more than one set of accelerations. He found $\dot{n}_M$ only in the sense that the application of Eqs. (I.1) to all sets yields the same value of $\dot{n}_M$. It does not yield the same value of the standard deviation of $\dot{n}_M$. Different writers give their own versions of the standard deviation, and I have taken the same liberty.

[†] Newton does not give this result explicitly. It comes from comparing the independent solutions that he found by using various constraints.

[‡] The value -41 was adopted before I discovered and corrected an error of $0''.8/cy^2$ in the value of $\dot{n}_M$ for the approximate epoch -200. The labor of recalculating the parabola did not seem warranted.

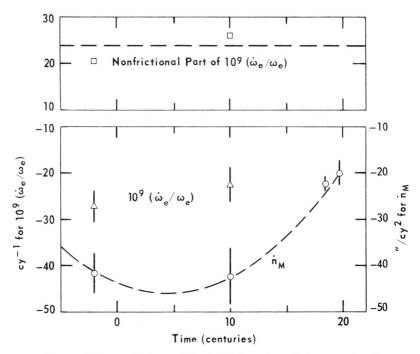

Figure XIV.1. Values of $\dot{n}_M$, $10^9 (\dot{\omega}_e/\omega_e)$, and the part of 10^9 $(\dot{\omega}_e/\omega_e)$ that is not attributable to tidal friction, plotted against time. The dashed horizontal line and the dashed curve are suggested time histories of the corresponding parameters.

$$\dot{n}_M = -22.0 \pm 1.1 + (3.3 \pm 1.2)T + (0.114 \pm 0.059)T^2 \; ''/cy^2$$

$$(XIV.12)$$

T, as usual, is time in centuries from 1900.

The values of $10^9(\dot{\omega}_e/\omega_e)$ are also plotted against their respective epochs in Figure XIV.1. The significance of the points and horizontal line marked "Nonfrictional part of $10^9(\dot{\omega}_e/\omega_e)$" will be discussed after a description of the second method of investigating the time history of the parameters.

The reader should now turn to Table XIV.1 and remember that $-A_i = y' = dy/dx$ for the lines in which $B_i = 1$. Eight lines in Table XIV.1 have y' near 0.6. In fact, y' is near this value, at least on the average, for any class of observation that involves the position of both the earth and moon, even if it involves the position of the earth only through a measurement of solar time.

Make the transformation of variable given by

$$Y = y - 0.622x .\qquad (XIV.13)$$

Six lines in Table XIV. 1 immediately furnish estimates of the variable Y at six different epochs from two different types of measurement. Two other lines furnish measurements of the variable $y - 0.576x$ at epochs near some of the six but from two more types of measurement. Now,

$$Y - (y - 0.576x) = -0.046x .$$

At the epochs of the combination $y - 0.576x$, x has been measured with an uncertainty that is probably no more than about $5''/cy^2$. Consequently, Y can be estimated from the measurements of $y - 0.576x$. The uncertainty in the resulting estimate of Y produced by the coordinate transformation should be no more than about $0''.2/cy^2$, which is small compared with the measurement error.

Six more estimates of Y can be found from the measurements in Table XI. 1 and in Tables XIII. 2 through XIII. 5. In order to form the estimates, consider Table XIII. 5, for example; the mean epoch of this table is about 1050. A matrix of the form shown in Eq. (XIV. 7) can be formed for the entries in this table. The matrix is nearly singular, reflecting the fact that the values of y' in the table are nearly the same. Consider the linear combination

$$-x \, [\Sigma a_i^2]^{\frac{1}{2}} + y \, [\Sigma b_i^2]^{\frac{1}{2}} \; .$$

This variable is strongly determined by the entries in the table. Any linearly independent variable is also determined, but weakly.

I formed a variable in this way for each of the tables cited in Chapter XIII for the large solar eclipses. I formed two variables for the solar eclipse magnitudes in

Table XI. 1, one for the Chinese data and one for the Islamic data. For the value of the measurement of each variable, I used the mean of $-(\Sigma z_i a_i)/[\Sigma a_i^2]^{\frac{1}{2}}$ and $+(\Sigma b_i z_i)/[\Sigma b_i^2]^{\frac{1}{2}}$. (The minus sign occurs here and earlier because the a_i are negative for all the measurements in question.) It was rare for the quantities that were averaged to differ by more than the standard deviation to be associated with the variable just defined. This standard deviation was estimated from the residuals of the z_i.

The variables found in this way, after re-normalization, were all nearly equal to Y and were transformed to it by the method already described. There are now 14 estimates of Y formed from six types of measurement for epochs fairly well distributed from -700 to $+1050$. Adding

TABLE XIV. 4

VALUES OF THE VARIABLE $10^9 (\dot{\omega}_e / \omega_e) - 0.622 \dot{n}_M$

Type of Measurement	Approx. Epoch	$10^9 (\dot{\omega}_e / \omega_e) - 0.622 \dot{n}_M, \ cy^{-1}$	
		Best Est.	Standard Dev.
Large solar eclipses before -400	-700	-1.0	0.2
Babylonian lunar eclipse times	-600	-2.4	0.7
Large solar eclipses between -400 & $+60$	-140	-1.9	0.2
Lunar occultations and conjunctions	-100	-3.0	1.1
Greek lunar eclipse times	-80	-0.6	0.7
Greek solar eclipse times	360	-5.1	1.8
Large solar eclipses between 60 & 500	420	-1.9	0.5
Chinese solar eclipse magnitudes	590	-5.1	4.2
Islamic lunar eclipse times before 950	900	$+4.4$	1.8
Islamic solar eclipse times	930	4.7	0.6
Islamic solar eclipse magnitudes	950	1.3	8.9
Islamic lunar eclipse times after 950	980	5.1	1.5
L_M from Hakemite tables	1000	3.5	0.4
Large solar eclipses after 500	1050	5.3	0.6

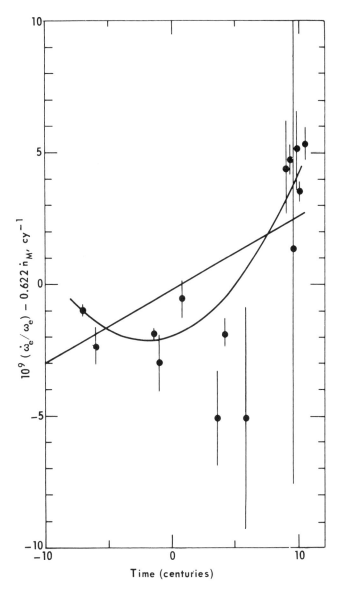

Figure XIV.2. Values of the parameter $10^9 (\dot{\omega}_e / \omega_e) - 0.622$ $\dot{n}_M$ plotted against time. The values come from six independent types of measurement. The straight line and the parabola were found by weighted least-squares. The figure suggests that the time variation was more complex than a quadratic.

14. 0 cy^{-1} to Y gives values of the variable $10^9(\dot{\omega}_e/\omega_e)$ - 0. 622$\dot{n}_M$. The estimates are plotted against time in Figure XIV. 2 and are listed in Table XIV. 4. The first column in the table describes the type of measurement, the second gives the approximate epoch, and the remaining two give the best estimate and the estimated standard deviation of the quantity tabulated.

There is a strong presumption that $10^9(\dot{\omega}_e/\omega_e)$ - 0. 622$\dot{n}_M$ has not been constant during historic times. A linear variation with time cannot be ruled out, but a more complex variation is far more likely. The straight line and the parabola in Figure XIV. 2 come from weighted least-squares fits. The weighted root-mean-square residual was reduced from 1. 6 to 0. 7 cy^{-1} in going from the straight line to the parabola. The probability that this could have happened by chance is less than 0. 01.

The equation of the parabola is

$$10^9(\dot{\omega}_e/\omega_e) - 0.622\dot{n}_M = -2.0 \pm 1.1 + (0.177 \pm 0.128)t$$
$$+ (0.0448 \pm 0.0219)t^2. \qquad \text{(XIV.14)}$$

If the error estimates are omitted, this is equivalent to

$$10^9(\dot{\omega}_e/\omega_e) - 0.622\dot{n}_M = -2.2 + 0.045(t + 1.98)^2 \quad \text{cy}^{-1}.$$

In these, t is time in centuries from 0.

The two values plotted in Figure XIV. 2 that came from measurements of solar eclipse magnitudes are the two with large standard deviations and with epochs near 600 and 950. Both lie on the same side of the parabola, and, at least for points in their general vicinity, have the largest residuals from the parabola. Thus they come from the only class of measurement for which a bias can be suggested with any degree of seriousness. However, the values plotted for epochs near 400 suggest that a "sawtooth" function is more appropriate than a parabola and hence that the value from magnitudes near 600 is not appreciably biased. I shall comment more on this possibility in Section XIV. 6.

Before taking up the possibility of a time variation of $\dot{\omega}_e$, it is desirable to discuss the geophysical sources of the accelerations briefly. The only known source of $\dot{n}_M$ large enough to need including in the present discussion is friction in the lunar tide. The same friction must be a source of $\dot{\omega}_e$. From our knowledge of moments of inertia of the earth and other relevant parameters, we can say with considerable accuracy that $10^9(\dot{\omega}_e/\omega_e) = 0.935\dot{n}_M$ so far as contributions from the lunar tide are concerned.

Friction in the solar tide contributes to $\dot{\omega}_e$ but not to $\dot{n}_M$. The deceleration of the sun (in this context, this means the deceleration of the earth in its orbit) that results from tidal friction is too small to measure. The only existing measurements of solar tidal friction have been made by means of satellite orbital data [Newton, 1968a; Kozai, 1967]. The results conflict with tidal theory, and it is not clear whether this implies defects in the theory or in the measurements. Since solar tidal friction is certainly much smaller than lunar tidal friction, it is sufficient here to use limits to the ratio of the two frictions from tidal theory.

Theories of tidal friction assume that almost all friction arises in the oceans. If friction is linear, the ratio of frictional torque in the lunar tide to that in the solar tide is close to the square of the ratio of the tide-raising forces; the torque ratio on this basis is about 5.1. Jeffreys [1962; pp. 231-235] showed that the torque ratio is reduced to about 3.8 if friction is quadratic[†] in the tidal velocities; he neglected the effects of oceanic motion of nontidal origin. Newton showed that the ratio is changed to the linear value 5.1 if the mean scalar velocity of nontidal

---

[†] Jeffreys evaluated the first two terms in a power series and found the value 3.4. This value has been used universally in the literature. Newton [1968a, p. 527] evaluated Jeffrey's function on a computer and found that the correct value is close to 3.8.

origin is more than about 1 cm/sec. We can do no better than to use the mean, 4.4 say, of the estimates of the ratio. On this basis the solar tide contributes $0.212\dot{n}_M$ to $10^9(\dot{\omega}_e/\omega_e)$ and the total frictional contribution is:

$$\text{frictional part of } 10^9(\dot{\omega}_e/\omega_e) = 1.147\dot{n}_M. \qquad \text{(XIV. 15)}$$

If the torque ratio lies between the theoretically derived limits, the coefficient in Eq. (XIV. 15) is in error by less than 4 percent.

Let us use y_{nf} to denote the part of $10^9(\dot{\omega}_e/\omega_e)$ that does not arise from tidal friction. Clearly,

$$y_{nf} = 10^9(\dot{\omega}_e/\omega_e) - 1.147\dot{n}_M \qquad \text{(XIV. 16)}$$

if Eq. (XIV. 15) is correct. The values of y_{nf} at the epochs -200 and +1000 are plotted near the top of Figure XIV.1, by the legend "Nonfrictional Part of $10^9(\dot{\omega}_e/\omega_e)$". For the purposes of investigating the geophysical sources of the accelerations, y_{nf} is more important than $\dot{\omega}_e$.

Equations (XIV. 12), (XIV. 14), and (XIV. 16) imply the following time dependence of y_{nf}:

$$y_{nf} = 20.7 + 0.72t - 0.0151t^2. \qquad \text{(XIV. 17)}$$

The coefficient of t^2 is not statistically significant, and the coefficient of t is marginally significant. For the present we can hardly do better than to speculate that y_{nf} is constant at the mean of the values in Figure XIV.1:

$$y_{nf} = 23 \pm 3 \text{ cy}^{-1}. \qquad \text{(XIV. 18)}$$

This agrees well with the constant part of Eq. (XIV. 17).

6. DISCUSSION; SOME GEOPHYSICAL SPECULATIONS

In all of the analysis in this study, except in part of Chapter IX, it has been assumed that the only uncertain parameters in the ephemerides of the sun-earth-moon system are the accelerations $\dot{n}_M$ and $\dot{\omega}_e$. It is pertinent to inquire whether the conclusions of the study might be modified if other parameters had been allowed to vary.

Positions of perigee are notoriously hard to compute, and we might first look for errors there. However, the positions of perigee of the solar and lunar orbits do not have an obvious large effect upon the observed quantities. The quantities that seem most likely to affect the measurements in a direct manner are the longitude Ω of the moon's node and the obliquity ϵ of the ecliptic.

It is probable that perturbations in Ω and ϵ could affect many individual observations by a significant amount. However, the averages over all eclipses, for example, in which the moon is sometimes north and sometimes south of the sun, sometimes near its ascending node and sometimes near its descending node, and in which the sun is sometimes north and sometimes south of the equator, should not be affected. Some types of observation, such as lunar occultations, should not be affected at all to first order. Thus perturbations in Ω and ϵ may apparently increase the scatter of the observations, but they should not directly affect the inferred accelerations.

The central line of a solar eclipse should be sensitive to perturbations in Ω. The possibility of a perturbation in Ω has caused me to abandon, at least in this study, the idea of trying to identify the "eclipse of Hipparchus".

A value of $\dot{n}_M$ as large in magnitude as $-42''/cy^2$ is hard to accommodate within existing geophysical theory. This value requires energy dissipation within the earth-moon system at the rate of 5.3×10^{19} ergs/sec while the value $\dot{n}_M = -22''.44/cy^2$ requires a rate of about 2.9×10^{19} ergs/sec. Taylor[†] in 1919 succeeded in calculating the

---

[†] The following discussion of energy dissipation is based upon the summary discussions by Jeffreys [1962, Chap. 8] and by Munk and MacDonald [1960, Chap. 11]. I have not consulted the original references cited by them and therefore do not give the citations, although I give names and

rate of dissipation in the Irish Sea. Within a few years Jeffreys and Heiskanen independently extended Taylor's method to all the waters. The basis of the method is the use of measured or estimated tidal velocities to estimate energy dissipation per unit area and then to integrate. There was agreement that most dissipation took place in shallow seas and that the Bering Sea alone accounted for a fractional part of the total dissipation ranging from 1/4 to 2/3, according to the estimate. Jeffreys found a dissipation rate of 1.1×10^{19} ergs/sec. Heiskanen's calculations, if we apply a correction to account for the fact that the measured currents that he used were for spring tides, lead to 1.9×10^{19} ergs/sec. Munk and MacDonald [1960, pp. 213-216] have reassessed the tidal currents used in the earlier calculations and have concluded that they were too high in some cases. Since the dissipation is assumed to vary with the cube of the velocity, the results are sensitive to the velocities used. Munk and MacDonald concluded that earlier estimates of dissipation in the Bering Sea, in particular, were high by perhaps an order of magnitude[†].

While the situation is still somewhat uncertain, it seems likely that the shallow seas, once considered to be the seat of most tidal dissipation, cannot contribute more than about $\frac{1}{2}$ of the amount required by the present value of $\dot{n}_M$.

Other sources, such as dissipation by flow over the continental shelves, the ocean floor, and at open beaches, have been studied and do not seem adequate, although there may be disagreement about whether the present value can be accounted for. Certainly no combination of mechanisms that I have seen discussed accounts for the ancient value needed.

---

dates in order to acknowledge the original work properly. The reader who wishes to consult the original references should have no trouble finding the citations.

[†] Miller [1966], in a later work, estimated that the dissipation in shallow seas is between 1.4 and 1.7×10^{19} ergs/sec. Apparently all the calculations mentioned ignore currents of nontidal origin.

It should be noticed that the problem does not seem to be whether the oceans dissipate energy at the rate required by the present value of $\dot{n}_M$. About 1921 Heiskanen calculated the rate at which energy is dissipated in the lunar tide by directly integrating the tide-raising potential over the oceans, using amplitudes from tidal charts to obtain the shape of the ocean surface as a function of time. He obtained a rate of 2.1×10^{19} ergs/sec for the lunar semi-diurnal tide. Groves and Munk repeated the calculation in 1959, using later tidal charts and extending the calculation to both lunar and solar tides and to both diurnal and semi-diurnal components. They obtained 3.2×10^{19} ergs/sec for the lunar tides. (The dissipation in the solar tides does not contribute to $\dot{n}_M$.) Since the dissipation rate is the difference between two large quantities, the final value could be considerably in error.

Thus the dissipation rates obtained from the current value of $\dot{n}_M$ and from direct integration of the tide-raising forces agree fairly well. The mechanisms that have been proposed do not seem to account for the observed dissipation, and at least one additional mechanism is needed.

If the results of this study, which require a change in dissipation by a factor of about two within historic times, be accepted, the magnitude of the change required may suggest the mechanism needed to account for the missing dissipation. It is hard to find a significant geophysical property that has changed greatly within historic times. Munk and MacDonald estimated the sensitivity of tidal friction to sea level, on the assumption that dissipation occurs near the coasts or in shallow seas, and concluded that the amount of friction might change by about one percent per meter change in sea level. Within the past two thousand years sea level may have changed by a few meters. Dissipation by the mechanisms that have been discussed, even if it could be made adequate in size, would not have changed by more than a few parts per hundred, far less than the factor of two needed.

The only relevant property I have thought of that has probably changed significantly within historic times is the amount of ice. Glaciers have been retreating rapidly [Brooks, 1958] for the past century or more. Variation in climate during the past two thousand years has not been uniform, however. From about 1200 to perhaps 1700 there was what has been called the "Little Ice Age". There was perhaps another full alternation of less and greater climatic severity between 1200 and 0. I do not know how the extent of glaciation in the year 0 compares with the present.

It is conceivable that fields of pack ice or shelf ice could be good absorbers of energy. Fields around Antarctica should be particularly important in the absorption of tidal energy because every major ocean is coupled directly to the coast of Antarctica or to the ice around it. If ice is an efficient absorber of energy, Antarctic ice would be in a favorable geographical position to absorb practically all the energy put into the tides by the tide-raising potential. Ice in the Arctic would have less opportunity to absorb tidal energy because it would not be well coupled to the major oceans, unless the ice extended well below the Aleutians and Greenland.

The large positive value of y_{nf}, the nonfrictional part of $10^9(\dot{\omega}_e/\omega_e)$, is also a source of embarassment. It has been considered for some time that y_{nf} is probably positive; I do not know who first stated this conclusion. The torque on the thermally induced atmospheric tide[†] contributes 2.7 to y_{nf}. Dicke [1964] has estimated that the overall effect of melting ice on the moment of inertia of the earth can give a contribution to y_{nf} lying between the limits +0.5 and +3.0, depending upon the rate of isostatic compensation assumed. Dicke [1964] and Newton [1968b] have used y_{nf} in attempts to discover time variations in the gravitational constant G. A thinkable rate of change of G could give a

---

[†] Holmberg [1952] calculated that the torque on the atmospheric tide is 3.7×10^{22} dyne-cm, which contributes 2.0 to y_{nf}. Newton [1968a] showed that the gravitational potential due to the tide in the atmosphere raises an earth tide. The torque on the combination of the atmospheric tide and the earth tide that it induces contributes 2.7 to y_{nf}.

contribution of the order of two or three to y_{nf}. From these processes, we can perhaps obtain a total of as much as nine, leaving us 14 short of the amount required by Eq. (XIV. 18)[†].

Cooling of the earth and consequent shrinking provides one possibility. The value $y_{nf} = 14$ would require that the moment of inertia of the earth be decreasing at the rate of 14 parts in 10^9 per century. It seems implausible to me that such a rate could have been maintained for as long as 10^7 cy $(10^9$ yr), but it is a superficially plausible rate to maintain for perhaps 10^6 cy. I gather from the literature that the idea of an already cooling and shrinking earth no longer finds the widespread acceptance that it once did.

We know that the earth's magnetic field undergoes considerable changes, even to the extent of reversing occasionally. If the magnetic field is accompanied by angular momentum of the core, something that we do not know, a change in the field would be accompanied by a change in the angular momentum of the core. Since a change in the field does not change the angular momentum of the total earth, there would be a consequent change in the angular momentum and the angular velocity of the solid parts of the earth where all ancient observers were situated. It is not safe at present to do more than speculate that the changes in the magnetic field that can be observed on an historic time scale can contribute to y_{nf}.

Munk and MacDonald [1960] summarized the calculations of a number of suggested contributions to y_{nf}, such as accretion of meteoric dust, and found none within an order of magnitude of the necessary amount.

[†]The weight of the bulge in the atmospheric tide loads the earth and oceans. The loading produces a torque of opposite sign to that exerted directly on the bulge. I have recently estimated ("Deceleration of the Earth's Spin Produced by the Atmospheric Tide", in preparation) that the total torque due to the atmospheric tide, to the tide induced by its gravitation, and to its loading tends to decelerate the earth's spin slightly. If this estimate is correct, the value 14 cy^{-1} in this discussion needs to be increased to about 17 cy^{-1}.

In returning to $\dot{n}_M$ briefly, I should point out that the time variation of $\dot{n}_M$ given by Eq. (XIV.12) does not tell the whole story. The ancient values of $\dot{n}_M$ that have been found are mean values between an ancient epoch and the epoch 1900, while the modern values can be considered as osculating values with regard to an historic time scale. Let $\nu_M(T)$ denote the osculating value of the acceleration of the moon. $\dot{n}_M$ and ν_M are related by

$$\dot{n}_M(T) = (2/T^2) \int_0^T dT_1 \int_0^{T_1} \nu_M(T_2)\, dT_2 .$$

In this, I have written T as an argument of $\dot{n}_M$ to emphasize that it varies with time. The solution of this for ν_M is

$$\nu_M = d^2(\tfrac{1}{2}T^2\dot{n}_M)/dT^2 . \qquad (XIV.19)$$

If Eq. (XIV.12) is correct, Eq. (XIV.19) leads to $\nu_M = -22 + 9.9T + 0.684T^2$. This function has the minimum value $\nu_M = -58''/cy^2$ compared with the minimum value $-46''/cy^2$ for $\dot{n}_M$. In other words, since $\dot{n}_M$ is an average, instantaneous values of the acceleration must have been even larger in magnitude than $\dot{n}_M$.

If $\dot{n}_M$ arises dominantly from friction, and no other possibility has been seriously advanced, the value just found for ν_M shows that Eq. (XIV.12) cannot possibly have been correct as far back as the epoch -200. The formula given is positive when T is outside the range from about -17 to about +2 cy, that is, from the epochs +200 to 2100. Thus Eq. (XIV.12) leads to a physical impossibility even within the time range to which it is meant to apply.

Therefore, the time history of $\dot{n}_M$ is probably more complex than Eq. (XIV.12) allows (unless it is in fact linear within historic times). This conclusion is also suggested by Figure XIV.2. The three values near the epoch 500 agree fairly well with each other and all are significantly below the parabola. The values in Figure XIV.2 in fact strongly suggest a sawtooth function, one that falls from 0 at about -1000 to about -4 at +500 and that then climbs abruptly to +5 at about +1000. Additional data for the time period from +500 to +1000 are urgently needed.

In summary, the time variations of the accelerations that can be found from the existing data can be represented well by

$$\dot{n}_M = -22 + 3.3T + 0.114T^2 \ ''/cy^2 ,$$

nonfrictional part of $10^9(\dot{\omega}_e/\omega_e) = +23 \ cy^{-1} ,$

$$(XIV.20)$$

for values of T between -21 and 0 cy. The actual time history is probably more complex than Eqs. (XIV.20) show, but we cannot infer more detail with confidence until we have more data[†].

If the second of Eqs. (XIV.20) has continued to hold up to the present, the secular acceleration of the earth's spin is now about zero and may even be positive. Whether this is so or not, the results of this study indicate that there have been times in the past when the nontidal contributions to $\dot{\omega}_e$ were greater than the tidal contributions are now. Thus it is not safe, even in approximate

---

[†]The results in Eqs. (XIV.20) were consistent with all known data at the time I first wrote Chapter XIV. However, I have just received (1970 Feb 27) some results from T. C. Van Flandern ("The Secular Acceleration of the Moon", in preparation), of the U.S. Naval Observatory, that may alter the situation. Van Flandern has analyzed occultations of stars by the moon between 1955 and 1969, using atomic clocks to furnish the time base. He finds the acceleration of the moon to be $-52 \pm 16 ''/cy^2$. This value agrees reasonably well with the ancient values found in this study, but it disagrees with Spencer Jones' result and with the results from artificial satellites. If the lunar acceleration has remained at or near the large values found from the ancient observations at the epochs -200 and +1000, the interpretation of Figure XIV.2 becomes more difficult but not impossible. The discrepancies with tidal theory remain.

calculations, to assume that the secular acceleration of the spin is largely governed by tidal friction. The large size of the nonfrictional part of $\dot{\omega}_e$ removes, at least temporarily, much of the basis for recent work by Dicke [1964] and Newton [1968b] on possible changes in the gravitational constant.

It is highly important to obtain additional data, particularly for the time period from 500 to 1000, and to see whether they will confirm or deny the main conclusions that have been reached in this study. I have obtained a large body of data for this time period, and I intend to analyze it as rapidly as possible. It seems undesirable to delay the publication of this work until these additional data can be analyzed.

This list is restricted to references that have been directly consulted in the course of preparing Part II of this study. Other works that come up in discussion, and that probably contain relevant information although they have not been directly consulted, are cited parenthetically or in footnotes.

Al-Battani, Astronomical Work, ca 925. There is a translation into Latin by Nallino, C. A., Publ. Reale Oss. di Brera, Milano, 1903.

American Ephemeris and Nautical Almanac, U.S. Government Printing Office, Washington, D. C. for various years.

Becvar, A., Atlas of the Heavens, Catalogue 1950. 0, Czechoslovak Acad. of Sciences, Prague, or Sky Publishing Corp., Cambridge, Mass., 1964.

Brooks, C. E. P., "Climate and Climatology," Encyclopedia Britannica, 5, Encyclopaedia Britannica, Inc., Chicago, 1958.

Curott, D. R., "Earth Deceleration from Ancient Solar Eclipses," Astron. J., 71, pp. 264-269, 1966.

Dicke, R. H., "The Secular Acceleration of the Earth's Rotation and Cosmology," paper presented at a Conference on the Earth-Moon System, Goddard Space Flight Center, January 1964; printed in The Earth-Moon System, B. G. Marsden and A. G. W. Cameron (eds.), Plenum Press, New York, pp. 98-164, 1966.

Ebn Iounis (or Ibn Junis), Le Livre de la Grande Table Hakémite, 1008. There is a translation into French by le C. en Caussin, Imprimerie de la République, Paris, 1804.

Eckert, W. J., Jones, Rebecca, and Clark, H. K., "Construction of the lunar ephemeris," in An Improved Lunar Ephemeris, 1952-1959, issued as a Joint Supplement to the American Ephemeris and Nautical Almanac and the (British) Nautical Almanac and Astronomical Ephemeris, U. S. Government Printing Office, Washington, D. C., 1954.

Explanatory Supplement to The Astronomical Ephemeris and The American Ephemeris and Nautical Almanac, H. M. Stationery Office, London, 1961.

Fotheringham, J. K., "On the Accuracy of the Alexandrian and Rhodian Eclipse Magnitudes," Monthly Not. Roy. Astro. Soc., 69, pp. 666-668, 1909.

_____, (assisted by Gertrude Longbottom), "The Secular Acceleration of the Moon's Mean Motion as Determined from the Occultations in the Almagest," Monthly Not. Roy. Astro. Soc., 75, pp. 377-394, 1915.

_____, "A Solution of Ancient Eclipses of the Sun," Monthly Not. Roy. Astro. Soc., 81, pp. 104-126, 1920.

Hill, G. W., "Tables of Jupiter, Constructed in Accordance with the Methods of Hansen," Astro. Papers Prepared for the Use of the Amer. Ephem. and Naut. Almanac, VII, Part 1, U. S. Government Printing Office, Washington, D. C., 1895a.

_____, "Tables of Saturn, Constructed in Accordance with the Methods of Hansen," Astro. Papers Prepared for the Use of the Amer. Ephem. and Naut. Almanac, VII, Part 2, U. S. Government Printing Office, Washington, D. C., 1895b.

Holmberg, E. R. R. , "A Suggested Explanation of the Present Value of the Velocity of Rotation of the Earth, " Monthly Not. Roy. Astro. Soc. , Geophys. Suppl. , 6, pp. 325-330, 1952.

Hydrographic Office (U. S.), Tables of Computed Altitude and Azimuth, Hydrographic Office Publication No. 214, U. S. Government Printing Office, Washington, D. C. , 1940.

Jeffreys, Sir Harold, The Earth, (4th ed.), Cambridge University Press, Cambridge, 1962.

Kozai, Y. , "Determination of Love's Number from Satellite Observations, " Trans. R. Soc. , A262, pp. 135-136, 1967.

Miller, G. R. , "The Flux of Tidal Energy out of the Deep Oceans, " J. Geophys. Res. , 71, pp. 2485-2489, 1966.

Munk, W. H. and MacDonald, G. J. F. , The Rotation of the Earth, Cambridge University Press, Cambridge, 1960.

Naval Observatory (U. S.), Ancient Sun and Moon, unpublished tables, 1953. The preparation of the tables is described by Woolard, E. W. , "Theory of the Rotation of the Earth Around its Center of Mass, " Astro. Papers Prepared for the Use of the Amer. Ephem. and Naut. Almanac, XV, Part I, U. S. Government Printing Office, Washington, D. C. , 1953.

Newcomb, S. , "Researches on the Motion of the Moon, " Washington Observations, U. S. Naval Observatory, Washington, 1875.

_____, "Tables of the Motion of the Earth on its Axis and around the Sun," Astro. Papers Prepared for the Use of the Amer. Ephem. and Naut. Almanac, VI, Part 1, U. S. Government Printing Office, Washington, D. C., 1895a.

_____, "Tables of Mercury," Astro. Papers Prepared for the Use of the Amer. Ephem. and Naut. Almanac, VI, Part 2, U. S. Government Printing Office, Washington, D. C., 1895b.

_____, "Tables of Venus," Astro. Papers Prepared for the Use of the Amer. Ephem. and Naut. Almanac, VI, Part 3, U. S. Government Printing Office, Washington, D. C., 1895c.

_____, "Tables of Mars," Astro. Papers Prepared for the Use of the Amer. Ephem. and Naut. Almanac, VI, Part 4, U. S. Government Printing Office, Washington, D. C., 1898.

Newton, R. R., "A Satellite Determination of Tidal Parameters and Earth Deceleration," Geophys. J. R. Astr. Soc., 14, pp. 505-539, 1968a.

_____, "Experimental Evidence for a Secular Decrease in the Gravitational Constant G," J. Geophys. Res., 73, pp. 3765-3771, 1968b.

Oppolzer, T. R. von, Canon der Finsternisse, Kaiserlich-Königlichen Hof- und Staatsdruckerei, Wien, 1887. Translation by O. Gingerich into English printed by Dover Publishing Co., New York, 1962.

Pannekoek, A., A History of Astronomy, Interscience Publishing Co., New York, 1961.

Ptolemy, C., 'E Mathematikè Syntaxis (The Almagest), ca
    152. This paper uses both the translation into French
    by M. Halma, Henri Grand Libraire, Paris, 1813,
    and the translation into German by K. Manitius,
    Leipzig (no publisher given), 1913. The chapter divi-
    sion and numbering is Halma's; there is an occasional
    difference of 1 from Manitius' chapter numbering.

Schroader, I. H., private communication, September 1968.

Spencer Jones, H., "The Rotation of the Earth, and the
    Secular Accelerations of the Sun, Moon, and Planets,"
    Monthly Not. Roy. Astro. Soc., 99, pp. 541-558,
    1939.

van der Waerden, B. L., Die Anfänge der Astronomie, P.
    Noordhoff Ltd., Groningen, The Netherlands, (Ger-
    man ed.), 1956.

_____, private communication, January 1969.

INDEX

A

Aberration
   planetary, 203
   stellar, 190, 203

Abydos, 97

Acceleration
   of earth's rotation
      defined, 2
      estimated, 34, 209, 271ff, 293
   of moon, estimated, 2, 34, 221,
      271ff, 293
   of "sun and moon", 1, 2, 5, 279
   time dependence of, 278ff, 293

Admont, Austria, 89

Agathocles, 103
   eclipse of, 103ff, 252

Agesilaos, 102ff

Agrippa, 162ff

Agrippina, 74

Airy, G. B., 97, 115

Al-Battani, 25, 149, 220, 235

Alessandria, Italy, 88

Alexandria, 10, 14, 17, 105ff, 115,
   128, 152, 154, 158, 159, 252,
   261ff

Alfred, King, 51

Altaich, Germany, 89

Altenzelle, Germany, 89

Alyattes, 96

Amajour (father and son), 165ff

Ammianus Marcellinus, 118

Amos, 61

Anglo-Saxon Chronicles, 50ff, 57,
   58, 60

Antioch, 146, 149, 220, 235

Appleby, J. T., 56ff

Archilochus, 99, 100, 114, 117
   eclipse of, 80, 91ff

Arezzo, Italy, 88

Aristarchus of Samos, 16, 116

Aristyllus, 23

Armenia, 73, 74, 113ff

ar-Raqqah, 24, 25, 146, 149, 220,
   235

Asia Minor, 95, 96, 99, 112

Assyria, 60, 61

Astyages, 96

Athens, 30, 101, 102, 106, 120,
   141

Atmospheric tide, 290, 291

Attouchement, 127, 150, 151

Augsburg, Germany, 89

B

Babylon, 57ff, 107, 135, 137,
   140ff, 235
   eclipse of, 58ff, 114

Bagdad, 24, 25, 28, 146, 164ff,
   173

Becvar, A., 156, 224

Bede, 49ff, 76ff

Beginning of eclipse
  defined, 124
  measurements of, Chap V.

Biot, M., 31

Bithynia, 111, 113, 158, 159

Boeotia, 102ff

Borissus, 119

Bright occultation, 160

Britton, J. P., 24, 138, 141, 142

Brooks, C. E. P., 290

C

Caesar, Julius, 44, 54, 82
  eclipse of, 70ff

Caifong-fu, 30

Cairo, 31ff, 146, 167ff, 173

Campania, Italy, 73, 74

Çanakkale, Turkey, 110

Canterbury, England, 51, 53, 55

Caussin, le C. en, 127

Celoria, G., 87ff

Cesena, Italy, 88

Chaeronia, Greece, 115

Chambers, G. F., 72

Chang: see Metonic cycle

Ch'ang-an: see Siganfou

Cherniss, H., 115ff

Chung K'ang, 62, 65, 121

Claudius, 74

Clemens, S. L., 45ff, 117

Cleombrotus, 98

Cleomedes, 104ff

Clerke, A. M., 21

Comet: see eclipse, comet seen
  during

Commodus, 74

Constantinople, 119, 121

Core of the earth, 291

Corinth, Isthmus of, 98

Corona: see eclipse, corona seen
  during

Cowell, P. H., 133

Croesus, 95

Curott, D. R., 3, 64, 69, 70, 104,
  251, 260

Cyaxares, 96

Cydias, 114

D

Damascus, 24, 25

Dandekar, B. S., 38

Dardanelles (town), 110

de Saussure, L., 64ff

de Sitter, W., 2, 3

Delphi, 115

Devizes, England, 56

Dicke, R. H., 290, 294

Dickins, B., 84

Digit, defined, 129

Digne, France, 88

Dio Cassius (or Dion Cassius),
70ff, 74, 78, 117, 118

Diodorus Siculus, 103ff

Dodgson, C. L., xvi

Dreyer, J. L. E., 114

Dubs, H. H., 43, 60, 62, 67, 69

Duncombe, R. L., 6, 187

E

Ebn Iounis, 25ff, 31ff, 127, 145ff,
155ff, 164ff, 200, 210

Ecclesiastical moon, 50, 51

Eckert, W. J., 187

Eclipse
assimilated, 98, 102, 118
defined, 43
comet seen during, 119
corona seen during, 38, 60,
114
cycles, 94, 95
literary, 92ff, 99, 102, 121
defined, 45
magical, 80, 81ff, 104, 111,
121
defined, 44
probability of noticing, 36ff
stars seen during, 37ff, 53, 54,
55, 69, 76, 100, 103, 111,
114, 118, 119, 260
see: beginning, end, magnitude,
middle, phases

Egypt, 61

Ellwangen, Germany, 89

End of eclipse
defined, 124
measurements of, Chap. V

Ensdorf, Germany, 89

Ephemeris longitude, 240ff, 246,
251ff

Ephemeris time, 2, 13, 139, 191,
200, 223, 225, 246, 278

Eponym canon eclipse, 60ff

Equal hours, defined, 26

Equation of time, 26

Equinoctial hours: see equal hours

Equinox, observations of, Chap. II,
268

Euctemon, 21, 29ff

Eusebius Pamphili, 110ff

F

Fang, 63ff

Firenze, Italy, 88

Fotheringham, J. K., throughout
Sect. IV.5; 2, 3, 5, 9ff, 18ff,
37, 58ff, 64, 126, 131, 133ff,
153ff, 156ff, 190ff, 212, 219,
222, 262, 275ff

France, 80

Frazer, J. G., 44, 65

Friction
in tides, 285ff, 294
in ice shelves, 290

Fuller, H. W., xvi

303

G

Gaubil, le Père, 29, 30, 62ff, 67ff, 143

Genoa, Italy, 88

Gervase of Canterbury, 57

Gilbert, W. S., 116

Gingerich, O., 24

Ginzel, F. K., 1, 40, 54, 61, 70ff, throughout Chap. IV

Gregory of Tours, 121

Groves, G., 289

Guier, W. H., xvi

H

Halley, E., 94

Halys River, 95ff

Han, Former, 30

Hangchou: see Hang-tcheou

Hang-tcheou, 30

Hayden, Fr. F. J., 36, 37

Heiskanen, W., 288, 289

Hellespont, 105ff, 252, 261ff

Henry
I of England, 52ff
II of England, 86
Prince, 86

Herodotus, 82, 94ff, 97ff, 108

Hill, G. W., 201ff

Hipparchus, 116, 142, 157
eclipse of, 104ff, 114, 128, 133ff, 153, 159, 243, 252, 257, 261ff, 273, 287
equinox observations, 9ff, 18, 27, 29, 157
length of year, 20
lunar parallax, 105ff
precession of equinoxes, 22
stellar observations, 10, 14

Hoang, P., 67, 68

Holland, A. W., 79

Holliday, C., xvi

Holmberg, E. R. R., 290

Holmes, Sherlock, 98

Holy Roman Empire, 89

Homer, 114

Hsi and Ho, 62ff

Hsiu, 64ff

Hungary, 90

Hydatius, 75, 76

Hydrographic Office (U. S.), 224

I

Iberian peninsula, 75, 119

Ibn Junis or Yunis: see Ebn Iounis

Identification game, xiv, 46, 93, 100, 105, 108, 260, 273

Interpretation factor, 158ff, 162ff, 199

Italy, 73, 113

J

Jebb, R. C., 101

304

Jeffreys, H., 214, 285, 287, 288

Jenkins, R. E., xvi

Jerome (St.), 110

Jesus, 111

Julian century, 4

Julian Ephemeris Date, 4

Julian Date, 4

Julius Obsequens, 70

Jupiter (planet), 166ff, 172, 201ff

K

Kaifeng, 69; see Caifong-fu

Kerulen River, 69ff

Knobel, E. G.: see Peters,
C. H. F.

Kongelbeck, S., xvi, 82

Kozai, Y., 285

Kratz, H., xvi, 82ff

Kugler, F. X., 61, 136, 138

L

Lambach, Austria, 89

Lammas Day, 52ff

Lampridius, 74

Landmark, J. D., 82ff

Larissa, eclipse of, 97, 121

Leo (constellation), 155, 165,
167ff, 172

Livy, 70, 78

Louis II of France, 52

Lu, state of, 69

Lucan, 73

Lucca, Italy, 88

Lydians, 94ff, 99

M

Macartney, C. A., 90

MacDonald, G. J. F.: see Munk,
W. H.

Macrinus, 117, 118

Magians, 97

Magnitude of eclipse, 36, 107ff,
123
accuracy of measurement, 131ff,
145, 217ff, 244
defined, 128
extended definition, 211, 239

Magnus V Erlingsson, 83, 86, 87

Magnus the Good, 84

Malmesbury, England, 54, 56

Marinus Neapolitanus, 119ff

Mars (planet), 164ff, 172, 201ff

Mauricius, 120, 121

Medes, 94ff, 97, 99

Mercury (planet), 164, 168ff, 201ff

Meton, 20, 29ff

Metonic cycle, 31, 66

Middle of eclipse, throughout
Chap. V
defined, 126

Milan, Italy, 88

Miller, G. R. , 288

Mimnermus, 114

Mirabeau, France, 88

Modena, Italy, 88

Mongolia (Outer), 69

Mongol invasion, 90

Montpelier, France, 88

Munk, W. H. , 3, 287ff

N

Nallino, C. A. , 25

Naval Observatory (U. S.), xvi, 12, 19, 25, 27, 187, 191, 202, 224

Needham, J. , 62ff, 68

Neresheim, Germany, 89

Nero, 74

Neugebauer, O. , 61, 94, 95, 137, 138, 142

Neugebauer, P. V. , 108

Neunburg, Austria, 89

Newcomb, S. , 1, 21, 52, 57, 97, 99, 115, 127, 135, 145ff, 153ff, 156, 187, 189, 201, 219

Newton, R. R. , 279, 285, 290, 294

Nicaea, 111ff

Nisabour, 24, 25

Node of lunar orbit, possible corrections needed, 218ff, 287

Noon, original meaning, 55

O

Obliquity of ecliptic, possible corrections needed, 287

Oenopides of Chios, 236

Olaf Haraldsson (St. Olaf), 81ff

O'Neill, M. J. , xvi

Oppolzer, T. R. von, 93
    Canon der Finsternisse, passim

Oslo, Norway, 86

P

Pannekoek, A. , 17, 22, 23, 25, 32, 66, 236

Pappus, 105ff

Parker, R. E. , xvi, 51, 53

Parma, Italy, 88

Paros, 92, 93

Peking, 30

Peloponnesian wars, 100, 102

Pericles, 101ff

Persia, 95, 97

Peterborough, England, 53, 55

Peters, C. H. F. , 18, 23

Phases of eclipses, 123ff, 227ff, 245ff, throughout Chap. V

Philostorgius, 118ff

Phlegon, 111ff
    eclipse of, 110ff, 121

Piacenza, Italy, 88

Pindar, 114, 117
  eclipse of, 99ff

Piraeus, 102

Pleiades, 155, 157, 162ff, 193ff

Pliny the Elder, 73, 94

Pliny the Younger, 73

Plummer, C., 51, 53

Plutarch, 44, 72, 73, 99, 101ff,
  103, 114ff
  eclipse of, 114ff, 121

Poland, 90

Pompey, 72ff, 78

Proclus, 83, 119ff

Ptolemy, 9, 16, 29, 61, 105ff,
  126, 128ff, 135, 152ff, 155ff,
  191, 197, 200, 216, 219, 235
  equinox and solstice observa-
    tions, 17ff
  errors in observations, 20ff
  king list, 60
  lunar observations, 23, 61
  precession of equinoxes, 22,
    32
  star catalog, 17, 23, 155

R

Rapp, A., xvi

Raqqa: see ar-Raqqah

Ravenna, Italy, 76ff, 88

Reggio, Italy, 88

Reliability of a record, 39, 158,
  264
  defined, 39

Rhodes, 9, 14, 152

Rich, R. P., xvi

Richard of Devizes, 56ff

Richersberg, Germany, 89

Roland, eclipse of, 78ff, 82, 260

Rome, 73, 74, 115, 158, 159

S

Sachs, A. J., 138

Salzburg, Austria, 89

Samarkand, 70, 144

Sardis, 97ff

Saros, 94

Sarton, G., 94

Saturn (planet), 164, 167, 168,
  171, 172, 201ff

Scheftlarn, Germany, 89

Schroader, I. H., xvi, 37, 209

Scorpio, 155, 157, 158, 160ff,
  166ff, 172, 192ff

Seasonal hours: see unequal hours

Shiraz, 165ff, 173

Sian: see Siganfou

Sicily, 93, 103

Siena, Italy, 88

Siganfou, 30, 143

Sign of zodiac
  as time unit, 32, 237ff
  use in keeping unequal hours,
    236ff

307

Sigvat the Scald, 81ff

Simocatta, Theophylactus, 120

Sin (the hsiu), 65

Smith, S., 59

Smithsonian Astronomical Observatory, 156

Snorri (Sturluson), 81ff

Solar time, 2, 4, 5, 12, 139, 191,
200, 223ff, 246

Solstice observations, 17ff, 24ff

Souciet, E., 69

Sparta, 93

Spencer Jones, H., 2, 3, 278, 293

Spica, 155, 157, 193ff

Stars seen during eclipse: see
eclipse, stars seen during

Stesichorus, 114

Stiklestad, eclipse of, 81ff

Storm, G., 86

Syncellus, 111

T

Taylor, G. I., 287, 288

Temporary hours: see unequal
hours

Teng-fong, 30

Thales, 94ff, 99
eclipse of, 7, 80, 94ff

Thasos, 92, 93, 101

Thebes, 100, 103

Theodosius, 119

Theon, eclipse of, 152, 154, 249

Thorarin Loftunga, 81

Thucydides, 37, 100ff
eclipse of, 100ff

Tigris River, 118

Time, accuracy of measuring, 127,
142ff, 147, 160, 237

Timocharis, 23, 162

Triple conjunction of planets, 167

Turoldus, 78

U

Unequal hours, 26, 148, 173, 236ff
measurement by stellar observation, 236ff

Universal time, 4, 5, 13

V

van der Waerden, B. L., 3, 106,
138, 141, 238, 274

van Flandern, T. C., 293

Venus (planet)
visibility during eclipse, 37
conjunctions and occultations,
164ff, 201ff

Vesuvius, 73

Vienna, Austria, 89

Virgil, 54, 73

W

Water clock, 142, 236, 248

Weedon, K., xvi, 87

Weihenstephan, Germany, 89

Weston, Jessie L., 44

Whitelock, D., 50, 51, 52, 55, 58

William of Malmesbury, 53ff

Winchester, England, 51, 56, 57

Wylie, A., 62, 67, 68, 69ff, 144

X

Xenophon, 97, 102

Xerxes, 97ff

Y

Yenching: see Teng-fong